D1766979

LIVERPOOL JMU LIBRARY

3  1111  01445  1783

# Ordinary Differential Equations

## Qualitative Theory

# Ordinary Differential Equations

## Qualitative Theory

Luis Barreira

Claudia Valls

Translated by the authors

Graduate Studies
in Mathematics

Volume 137

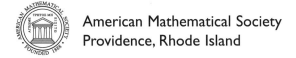

American Mathematical Society
Providence, Rhode Island

# EDITORIAL COMMITTEE

David Cox (Chair)
Daniel S. Freed
Rafe Mazzeo
Gigliola Staffilani

This work was originally published in Portuguese by IST Press under the title "Equações Diferenciais: Teoria Qualitativa" by Luis Barreira and Clàudia Valls, © IST Press 2010, Instituto Superior Técnico. All Rights Reserved.

The present translation was created under license for the American Mathematical Society and published by permission.

Translated by the authors.

2010 *Mathematics Subject Classification.* Primary 34-01, 34Cxx, 34Dxx, 37Gxx, 37Jxx.

For additional information and updates on this book, visit
**www.ams.org/bookpages/gsm-137**

**Library of Congress Cataloging-in-Publication Data**

Barreira, Luis, 1968–
  [Equações diferenciais. English]
  Ordinary differential equations : qualitative theory / Luis Barreira, Claudia Valls ; translated by the authors.
      p. cm. – (Graduate studies in mathematics ; v. 137)
  Includes bibliographical references and index.
  ISBN 978-0-8218-8749-3 (alk. paper)
  1. Differential equations–Qualitative theory.  I. Valls, Claudia, 1973–  II. Title.

QA372.B31513    2010
515′.352–dc23

2012010848

**Copying and reprinting.** Individual readers of this publication, and nonprofit libraries acting for them, are permitted to make fair use of the material, such as to copy a chapter for use in teaching or research. Permission is granted to quote brief passages from this publication in reviews, provided the customary acknowledgment of the source is given.

Republication, systematic copying, or multiple reproduction of any material in this publication is permitted only under license from the American Mathematical Society. Requests for such permission should be addressed to the Acquisitions Department, American Mathematical Society, 201 Charles Street, Providence, Rhode Island 02904-2294 USA. Requests can also be made by e-mail to reprint-permission@ams.org.

© 2012 by the American Mathematical Society. All rights reserved.
The American Mathematical Society retains all rights
except those granted to the United States Government.
Printed in the United States of America.

♾ The paper used in this book is acid-free and falls within the guidelines
established to ensure permanence and durability.
Visit the AMS home page at http://www.ams.org/

10 9 8 7 6 5 4 3 2 1      17 16 15 14 13 12

# Contents

# Preface

The main objective of this book is to give a comprehensive introduction to the qualitative theory of ordinary differential equations. In particular, among other topics, we study the existence and uniqueness of solutions, phase portraits, linear equations and their perturbations, stability and Lyapunov functions, hyperbolicity, and equations in the plane.

The book is also intended to serve as a bridge to important topics that are often left out of a second course of ordinary differential equations. Examples include the smooth dependence of solutions on the initial conditions, the existence of topological and differentiable conjugacies between linear systems, and the Hölder continuity of the conjugacies in the Grobman–Hartman theorem. We also give a brief introduction to bifurcation theory, center manifolds, normal forms, and Hamiltonian systems.

We describe mainly notions, results and methods that allow one to discuss the qualitative properties of the solutions of an equation without solving it explicitly. This can be considered the main aim of the qualitative theory of ordinary differential equations.

The book can be used as a basis for a second course of ordinary differential equations. Nevertheless, it has more material than the standard courses, and so, in fact, it can be used in several different ways and at various levels. Among other possibilities, we suggest the following courses:

a) advanced undergraduate/beginning graduate second course: Chapters 1–5 and 7–8 (without Sections 1.4, 2.5 and 8.3, and without the proofs of the Grobman–Hartman and Hadamard–Perron theorems);

b) advanced undergraduate/beginning graduate course on equations in the plane: Chapters 1–3 and 6–7;

c) advanced graduate course on stability: Chapters 1–3 and 8–9;

d) advanced graduate course on hyperbolicity: Chapters 1–5.

Other selections are also possible, depending on the audience and on the time available for the course. In addition, some sections can be used for short expositions, such as Sections 1.3.2, 1.4, 2.5, 3.3, 6.2 and 8.3.

Other than some basic pre-requisites of linear algebra and differential and integral calculus, all concepts and results used in the book are recalled along the way. Moreover, (almost) everything is proven, with the exception of some results in Chapters 8 and 9 concerning more advanced topics of bifurcation theory, center manifolds, normal forms and Hamiltonian systems. Being self-contained, the book can also serve as a reference or for independent study.

Now we give a more detailed description of the contents of the book. Part 1 is dedicated to basic concepts and linear equations.

- In Chapter 1 we introduce the basic notions and results of the theory of ordinary differential equations, in particular, concerning the existence and uniqueness of solutions (Picard–Lindelöf theorem) and the dependence of solutions on the initial conditions. We also establish the existence of solutions of equations with a continuous vector field (Peano's theorem). Finally, we give an introduction to the description of the qualitative behavior of the solutions in the phase space.

- In Chapter 2 we consider the particular case of (nonautonomous) linear equations and we study their fundamental solutions. It is often useful to see an equation as a perturbation of a linear equation, and to obtain the solutions (even if implicitly) using the variation of parameters formula. This point of view is often used in the book. We then consider the particular cases of equations with constant coefficients and equations with periodic coefficients. More advanced topics include the $C^1$ dependence of solutions on the initial conditions and the existence of topological conjugacies between linear equations with hyperbolic matrices of coefficients.

Part 2 is dedicated to the study of stability and hyperbolicity.

- In Chapter 3, after introducing the notions of stability and asymptotic stability, we consider the particular case of linear equations, for which it is possible to give a complete characterization of these notions in terms of fundamental solutions. We also consider the particular cases of equations with constant coefficients and equations with periodic coefficients. We then discuss the persistence of asymptotic stability under sufficiently small perturbations of an asymptotically stable linear equation. We also give an introduction to the theory of Lyapunov functions, which sometimes yields the stability of a given solution in a more or less automatic manner.

- In Chapters 4–5 we introduce the notion of hyperbolicity and we study some of its consequences. Namely, we establish two key results on the behavior of the solutions in a neighborhood of a hyperbolic critical point: the Grobman–Hartman and Hadamard–Perron theorems. The first shows that the solutions of a sufficiently small perturbation of a linear equation with a hyperbolic critical point are topologically conjugate to the solutions of the linear equation. The second shows that there are invariant manifolds tangent to the stable and unstable spaces of a hyperbolic critical point. As a more advanced topic, we show that all conjugacies in the Grobman–Hartman theorem are Hölder continuous. We note that Chapter 5 is more technical: the exposition is dedicated almost entirely to the proof of the Hadamard–Perron theorem. In contrast to what happens in other texts, our proof does not require a discretization of the problem or additional techniques that would only be used here. We note that the material in Sections 5.3 and 5.4 is used nowhere else in the book.

In Part 3 we describe results and methods that are particularly useful in the study of equations in the plane.

- In Chapter 6 we give an introduction to index theory and its applications to differential equations in the plane. In particular, we describe how the index of a closed path with respect to a vector field varies with the path and with the vector field. We then present several applications, including a proof of the existence of a critical point inside any periodic orbit, in the sense of Jordan's curve theorem.

- In Chapter 7 we give an introduction to the Poincaré–Bendixson theory. After introducing the notions of $\alpha$-limit and $\omega$-limit sets, we show that bounded semiorbits have nonempty, compact and connected $\alpha$-limit and $\omega$-limit sets. Then we establish one of the important results of the qualitative theory of ordinary differential equations in the plane, the Poincaré–Bendixson theorem. In particular, it yields a criterion for the existence of periodic orbits.

Part 4 is of a somewhat different nature and it is only here that not everything is proved. Our main aim is to make the bridge to important topics that are often left out of a second course of ordinary differential equations.

- In Chapter 8 we give an introduction to bifurcation theory, with emphasis on examples. We then give an introduction to the theory of center manifolds, which often allows us to reduce the order of an equation in the study of stability or the existence of bifurcations. We also give an introduction to the theory of normal forms that aims to eliminate through a change of variables all possible terms in the original equation.

- Finally, in Chapter 9 we give an introduction to the theory of Hamiltonian systems. After introducing some basic notions, we describe several results concerning the stability of linear and nonlinear Hamiltonian systems. We also consider the notion of integrability and the Liouville–Arnold theorem on the structure of the level sets of independent integrals in involution. In addition, we describe the basic ideas of the KAM theory.

The book also includes numerous examples that illustrate in detail the new concepts and results as well as exercises at the end of each chapter.

Luis Barreira and Claudia Valls

Lisbon, February 2012

*Part 1*

# Basic Concepts and Linear Equations

# Chapter 1

# Ordinary Differential Equations

In this chapter we introduce the basic notions of the theory of ordinary differential equations, including the concepts of solution and of initial value problem. We also discuss the existence and uniqueness of solutions, their dependence on the initial conditions, and their behavior at the endpoints of the maximal interval of existence. To that effect, we recall the relevant material concerning contractions and fiber contractions. Moreover, we show how the solutions of an autonomous equation may give rise to a flow. Finally, we give an introduction to the qualitative theory of differential equations, with the discussion of many equations and of their phase portraits. We include in this study the particular case of the conservative equations, that is, the equations with an integral. For additional topics we refer the reader to [**2, 9, 13, 15**].

## 1.1. Basic notions

In this section we introduce the notions of a solution of an ordinary differential equation and of initial value problem, and we illustrate them with various examples. We also show how the solutions of an autonomous equation (that is, an equation not depending explicitly on time) may give rise to a flow.

**1.1.1. Solutions and initial value problems.** We first introduce the notion of a solution of an ordinary differential equation. Given a continuous function $f\colon D \to \mathbb{R}^n$ in an open set $D \subset \mathbb{R} \times \mathbb{R}^n$, consider the *(ordinary)* *differential equation*

$$x' = f(t, x). \tag{1.1}$$

The unknown of this equation is the function $x = x(t)$.

**Definition 1.1.** A function $x \colon (a, b) \to \mathbb{R}^n$ of class $C^1$ (with $a \geq -\infty$ and $b \leq +\infty$) is said to be a *solution* of equation (1.1) if (see Figure 1.1):

    a) $(t, x(t)) \in D$ for every $t \in (a, b)$;

    b) $x'(t) = f(t, x(t))$ for every $t \in (a, b)$.

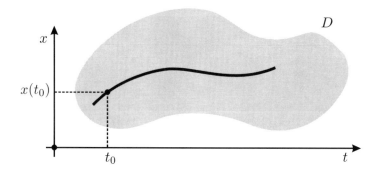

**Figure 1.1.** A solution $x = x(t)$ of the equation $x' = f(t, x)$.

**Example 1.2.** Consider the equation

$$x' = -x + t. \tag{1.2}$$

If $x = x(t)$ is a solution, then

$$(e^t x)' = e^t x + e^t x' = e^t(x + x') = e^t t.$$

Since a primitive of $e^t t$ is $e^t(t - 1)$, we obtain

$$e^t x(t) = e^t(t - 1) + c$$

for some constant $c \in \mathbb{R}$. Therefore, the solutions of equation (1.2) are given by

$$x(t) = t - 1 + ce^{-t}, \quad t \in \mathbb{R}, \tag{1.3}$$

with $c \in \mathbb{R}$.

Now we consider an equation in $\mathbb{R}^2$.

**Example 1.3.** Consider the equation

$$(x, y)' = (y, -x), \tag{1.4}$$

which can be written in the form

$$\begin{cases} x' = y, \\ y' = -x. \end{cases} \tag{1.5}$$

If $(x(t), y(t))$ is a solution, then

$$(x^2 + y^2)' = 2xx' + 2yy' = 2xy + 2y(-x) = 0. \tag{1.6}$$

Thus, there exists a constant $r \geq 0$ such that

$$x(t)^2 + y(t)^2 = r^2$$

(for every $t$ in the domain of the solution). Writing

$$x(t) = r\cos\theta(t) \quad \text{and} \quad y(t) = r\sin\theta(t),$$

where $\theta$ is a differentiable function, it follows from the identity $x' = y$ that

$$x'(t) = -r\theta'(t)\sin\theta(t) = r\sin\theta(t).$$

Hence, $\theta'(t) = -1$, and there exists a constant $c \in \mathbb{R}$ such that $\theta(t) = -t + c$. Therefore,

$$(x(t), y(t)) = \big(r\cos(-t+c), r\sin(-t+c)\big), \quad t \in \mathbb{R}, \tag{1.7}$$

with $c \in \mathbb{R}$. These are all the solutions of equation (1.4).

**Example 1.4.** It follows from (1.5) that

$$x'' = (x')' = y' = -x.$$

On the other hand, starting with the equation $x'' = -x$ and writing $y = x'$, we obtain $y' = x'' = -x$. In other words, we recover the two identities in (1.5). Hence, the equations (1.5) and $x'' = -x$ are equivalent.

More generally, writing

$$X = (X_1, \ldots, X_k) = (x, x', \ldots, x^{(k-1)}),$$

the equation

$$x^{(k)} = f(t, x, x', \ldots, x^{(k-1)})$$

can be written in the form $X' = F(t, X)$, where

$$F(t, X) = \big(X_2, X_3, \ldots, X_{k-1}, f(t, X)\big).$$

One can also consider differential equations written in other coordinates, such as, for example, in polar coordinates.

**Example 1.5.** Consider the equation

$$\begin{cases} r' = r, \\ \theta' = 1 \end{cases} \tag{1.8}$$

written in polar coordinates $(r, \theta)$. Since $x = r\cos\theta$ and $y = r\sin\theta$, we obtain

$$x' = r'\cos\theta - r\theta'\sin\theta = r\cos\theta - r\sin\theta = x - y \tag{1.9}$$

and

$$y' = r'\sin\theta + r\theta'\cos\theta = r\sin\theta + r\cos\theta = y + x. \tag{1.10}$$

Thus, equation (1.8) can be written in the form

$$\begin{cases} x' = x - y, \\ y' = x + y. \end{cases} \tag{1.11}$$

Incidentally, one can solve separately each equation in (1.8) to obtain the solutions

$$\begin{cases} r(t) = ce^t, \\ \theta(t) = t + d, \end{cases} \tag{1.12}$$

where $c, d \in \mathbb{R}$ are arbitrary constants.

Now we introduce the notion of initial value problem.

**Definition 1.6.** Given $(t_0, x_0) \in D$, the *initial value problem*

$$\begin{cases} x' = f(t, x), \\ x(t_0) = x_0 \end{cases} \tag{1.13}$$

consists of finding an interval $(a, b)$ containing $t_0$ and a solution $x \colon (a, b) \to \mathbb{R}^n$ of equation (1.1) such that $x(t_0) = x_0$. The condition $x(t_0) = x_0$ is called the *initial condition* of problem (1.13).

**Example 1.7** (continuation of Example 1.2). Consider the initial value problem

$$\begin{cases} x' = -x + t, \\ x(0) = 0. \end{cases} \tag{1.14}$$

Taking $t = 0$ in (1.3), we obtain $x(0) = -1 + c$, and the initial condition yields $c = 1$. Hence, a solution of the initial value problem (1.14) is

$$x(t) = t - 1 + e^{-t}, \quad t \in \mathbb{R}.$$

**Example 1.8** (continuation of Example 1.3). Consider equation (1.4) and the corresponding initial value problem

$$\begin{cases} (x, y)' = (y, -x), \\ (x(t_0), y(t_0)) = (x_0, y_0), \end{cases} \tag{1.15}$$

where $t_0 \in \mathbb{R}$ and $(x_0, y_0) \in \mathbb{R}^2$. Taking $t = t_0$ in (1.7), we obtain

$$(x_0, y_0) = \big(r \cos(-t_0 + c), r \sin(-t_0 + c)\big).$$

Writing the initial condition in polar coordinates, that is,

$$(x_0, y_0) = (r \cos \theta_0, r \sin \theta_0), \tag{1.16}$$

we obtain $c = t_0 + \theta_0$ (up to an integer multiple of $2\pi$). Hence, a solution of the initial value problem (1.15) is

$$(x(t), y(t)) = \big(r \cos(-t + t_0 + \theta_0), r \sin(-t + t_0 + \theta_0)\big), \quad t \in \mathbb{R}. \tag{1.17}$$

An equivalent and more geometric description of the solution (1.17) of problem (1.15) is the following. By (1.16) and (1.17), we have

$$x(t) = r\cos\theta_0\cos(-t + t_0) - r\sin\theta_0\sin(-t + t_0)$$
$$= x_0\cos(t - t_0) + y_0\sin(t - t_0),$$

and analogously (or simply observing that $y = x'$),

$$y(t) = -x_0\sin(t - t_0) + y_0\cos(t - t_0).$$

Thus, one can write the solution (1.17) in the form

$$\begin{pmatrix} x(t) \\ y(t) \end{pmatrix} = R(t - t_0)\begin{pmatrix} x_0 \\ y_0 \end{pmatrix}, \tag{1.18}$$

where

$$R(t) = \begin{pmatrix} \cos t & \sin t \\ -\sin t & \cos t \end{pmatrix}$$

is a rotation matrix. Indeed, as shown in Example 1.3, each solution of equation (1.4) remains at a fixed distance from the origin.

The following is a characterization of the solutions of an ordinary differential equation.

**Proposition 1.9.** *Let $f\colon D \to \mathbb{R}^n$ be a continuous function in an open set $D \subset \mathbb{R} \times \mathbb{R}^n$. Given $(t_0, x_0) \in D$, a continuous function $x\colon (a, b) \to \mathbb{R}^n$ in an open interval $(a, b)$ containing $t_0$ is a solution of the initial value problem (1.13) if and only if*

$$x(t) = x_0 + \int_{t_0}^t f(s, x(s))\,ds \tag{1.19}$$

*for every $t \in (a, b)$.*

**Proof.** We first note that the function $t \mapsto f(t, x(t))$ is continuous, because it is a composition of continuous functions. In particular, it is also integrable in each bounded interval. Now let us assume that $x = x(t)$ is a solution of the initial value problem (1.13). For each $t \in (a, b)$, we have

$$x(t) - x_0 = x(t) - x(t_0) = \int_{t_0}^t x'(s)\,ds = \int_{t_0}^t f(s, x(s))\,ds,$$

which establishes (1.19). On the other hand, if identity (1.19) holds for every $t \in (a, b)$, then clearly $x(t_0) = x_0$, and taking derivatives with respect to $t$, we obtain

$$x'(t) = f(t, x(t)) \tag{1.20}$$

for every $t \in (a, b)$. Since the function $t \mapsto f(t, x(t))$ is continuous, it follows from (1.20) that $x$ is of class $C^1$, and hence it is a solution of the initial value problem (1.13). □

**1.1.2. Notion of a flow.** In this section we consider the particular case of ordinary differential equations not depending explicitly on time, and we show that they naturally give rise to the notion of a flow.

**Definition 1.10.** Equation (1.1) is said to be *autonomous* if $f$ does not depend on $t$.

In other words, an autonomous equation takes the form

$$x' = f(x),$$

where $f \colon D \to \mathbb{R}^n$ is a continuous function in an open set $D \subset \mathbb{R}^n$.

Now we introduce the notion of a flow.

**Definition 1.11.** A family of transformations $\varphi_t \colon \mathbb{R}^n \to \mathbb{R}^n$ for $t \in \mathbb{R}$ such that $\varphi_0 = \mathrm{Id}$ and

$$\varphi_{t+s} = \varphi_t \circ \varphi_s \quad \text{for} \quad t, s \in \mathbb{R} \tag{1.21}$$

is called a *flow*.

**Example 1.12.** Given $y \in \mathbb{R}^n$, the family of transformations $\varphi_t \colon \mathbb{R}^n \to \mathbb{R}^n$ defined by

$$\varphi_t(x) = x + ty, \quad t \in \mathbb{R}, \ x \in \mathbb{R}^n$$

is a flow.

The following statement shows that many autonomous equations give rise to a flow.

**Proposition 1.13.** *If $f \colon \mathbb{R}^n \to \mathbb{R}^n$ is a continuous function such that each initial value problem*

$$\begin{cases} x' = f(x), \\ x(0) = x_0 \end{cases} \tag{1.22}$$

*has a unique solution $x(t, x_0)$ defined for $t \in \mathbb{R}$, then the family of transformations defined by*

$$\varphi_t(x_0) = x(t, x_0) \tag{1.23}$$

*is a flow.*

**Proof.** Given $s \in \mathbb{R}$, consider the function $y \colon \mathbb{R} \to \mathbb{R}^n$ defined by

$$y(t) = x(t + s, x_0).$$

Clearly, $y(0) = x(s, x_0)$ and

$$y'(t) = x'(t + s, x_0) = f(x(t + s, x_0)) = f(y(t))$$

for $t \in \mathbb{R}$. In particular, $y$ is also a solution of the equation $x' = f(x)$. By hypothesis each initial value problem (1.22) has a unique solution, and hence,

$$y(t) = x(t, y(0)) = x(t, x(s, x_0)),$$

or equivalently,

$$x(t + s, x_0) = x(t, x(s, x_0)), \tag{1.24}$$

for every $t, s \in \mathbb{R}$ and $x_0 \in \mathbb{R}^n$. Using the transformations $\varphi_t$ in (1.23), one can rewrite identity (1.24) in the form

$$\varphi_{t+s}(x_0) = (\varphi_t \circ \varphi_s)(x_0).$$

On the other hand, $\varphi_0(x_0) = x(0, x_0) = x_0$, by the definition of $x(t, x_0)$. This shows that the family of transformations $\varphi_t$ is a flow. □

**Example 1.14** (continuation of Examples 1.3 and 1.8). We note that equation (1.4) is autonomous. Moreover, the initial value problem (1.15) has a unique solution that is defined in $\mathbb{R}$, given by (1.17) or (1.18). Thus, it follows from Proposition 1.13 that the family of transformations $\varphi_t$ defined by

$$\varphi_t(x_0, y_0) = \big(r\cos(-t + \theta_0), r\sin(-t + \theta_0)\big)$$

is a flow. Using (1.18), one can also write

$$\varphi_t(x_0, y_0) = (x(t), y(t)) = (x_0, y_0)R(t)^*,$$

where $R(t)^*$ is the transpose of the rotation matrix $R(t)$. Hence, identity (1.21) is equivalent to

$$R(t + s)^* = R(s)^* R(t)^*,$$

or taking transposes,

$$R(t + s) = R(t)R(s),$$

for $t, s \in \mathbb{R}$. This is a well-known identity between rotation matrices.

**Example 1.15** (continuation of Example 1.5). Taking $t = 0$ in (1.12), we obtain $r(0) = c$ and $\theta(0) = d$. Thus, the solutions of equation (1.8) can be written (in polar coordinates) in the form

$$(r(t), \theta(t)) = (r_0 e^t, t + \theta_0),$$

where $(r_0, \theta_0) = (r(0), \theta(0))$. If $\varphi_t$ is the flow given by Proposition 1.13 for the autonomous equation (1.11), then in coordinates $(x, y)$ the point $(r(t), \theta(t))$ can be written in the form

$$\varphi_t(r_0 \cos\theta_0, r_0 \sin\theta_0) = \big(r(t)\cos\theta(t), r(t)\sin\theta(t)\big)$$
$$= \big(r_0 e^t \cos(t + \theta_0), r_0 e^t \sin(t + \theta_0)\big).$$

## 1.2. Existence and uniqueness of solutions

In this section we discuss the existence and uniqueness of solutions of an ordinary differential equation. After formulating the results, we make a brief digression on auxiliary material that is used in the proofs of these and of other results in the book.

**1.2.1. Formulation of the Picard–Lindelöf theorem.** We first recall the notion of a compact set.

**Definition 1.16.** A set $K \subset \mathbb{R}^m$ is said to be *compact* if given an open cover of $K$, that is, a family of open sets $(U_\alpha)_{\alpha \in I} \subset \mathbb{R}^m$ such that $\bigcup_{\alpha \in I} U_\alpha \supset K$, there exists a finite subcover, that is, there exist $N \in \mathbb{N}$ and $\alpha_1, \ldots, \alpha_N \in I$ such that $\bigcup_{i=1}^N U_{\alpha_i} \supset K$.

We recall that a set $K \subset \mathbb{R}^m$ is compact if and only if it is closed and bounded.

Now we introduce the notion of a locally Lipschitz function.

**Definition 1.17.** A function $f \colon D \to \mathbb{R}^n$ in a set $D \subset \mathbb{R} \times \mathbb{R}^n$ is said to be *locally Lipschitz* in $x$ if for each compact set $K \subset D$ there exists $L > 0$ such that

$$\|f(t, x) - f(t, y)\| \le L\|x - y\| \tag{1.25}$$

for every $(t, x), (t, y) \in K$.

Using this notion, one can formulate the following result on the existence and uniqueness of solutions of an ordinary differential equation.

**Theorem 1.18** (Picard–Lindelöf theorem). *If the function $f \colon D \to \mathbb{R}^n$ is continuous and locally Lipschitz in $x$ in an open set $D \subset \mathbb{R} \times \mathbb{R}^n$, then for each $(t_0, x_0) \in D$ there exists a unique solution of the initial value problem* (1.13) *in some open interval containing $t_0$.*

The proof of Theorem 1.18 is given in Section 1.2.3, using some of the auxiliary results described in Section 1.2.2.

In particular, all $C^1$ functions are locally Lipschitz.

**Proposition 1.19.** *If $f$ is of class $C^1$, then $f$ is locally Lipschitz in $x$.*

**Proof.** We first show that it is sufficient to consider the family $\mathcal{K}$ of compact sets of the form $I \times J \subset D$, where $I \subset \mathbb{R}$ is a compact interval and

$$J = \big\{ x \in \mathbb{R}^n : \|x - p\| \le r \big\}$$

for some $p \in \mathbb{R}^n$ and $r > 0$. To that effect, we show that each compact set $K \subset D$ is contained in a finite union of elements of $\mathcal{K}$. Then, let $K \subset D$ be a compact set. For each $p \in K$, take $K_p \in \mathcal{K}$ such that $p \in \operatorname{int} K_p$. Clearly, $K \subset \bigcup_{p \in K} \operatorname{int} K_p$, and thus, it follows from the compactness of $K$ that there exist $N \in \mathbb{N}$ and $p_1, \ldots, p_N \in K$ such that $K \subset \bigcup_{i=1}^N \operatorname{int} K_{p_i}$. Hence, $K \subset \bigcup_{i=1}^N K_{p_i} \subset D$, which yields the desired result.

Now let $I \times J$ be a compact set in $\mathcal{K}$ and take points $(t, x), (t, y) \in I \times J$. The function $g \colon [0, 1] \to D$ defined by

$$g(s) = f(t, y + s(x - y))$$

is of class $C^1$, because it is a composition of $C^1$ functions. Moreover,

$$f(t,x) - f(t,y) = g(1) - g(0) = \int_0^1 g'(s)\, ds$$

$$= \int_0^1 \frac{\partial f}{\partial x}(t, y + s(x-y))(x-y)\, ds,$$

where $\partial f / \partial x$ is the Jacobian matrix (thinking of the vector $x - y$ as a column). Thus,

$$\begin{aligned}
\|f(t,x) - f(t,y)\| &\leq \sup_{s \in [0,1]} \left\| \frac{\partial f}{\partial x}(t, y + s(x-y)) \right\| \cdot \|x - y\| \\
&\leq \sup_{q \in J} \left\| \frac{\partial f}{\partial x}(t, q) \right\| \cdot \|x - y\|,
\end{aligned} \tag{1.26}$$

because the line segment between $x$ and $y$ is contained in $J$. Now we observe that since $f$ is of class $C^1$, the function

$$(t, q) \mapsto \left\| \frac{\partial f}{\partial x}(t, q) \right\|$$

is continuous, and hence,

$$L := \max_{(t,q) \in I \times J} \left\| \frac{\partial f}{\partial x}(t, q) \right\| < +\infty.$$

It follows from (1.26) that

$$\|f(t,x) - f(t,y)\| \leq L \|x - y\|$$

for every $(t, x), (t, y) \in I \times J$. This shows that $f$ is locally Lipschitz in $x$. $\square$

The following is a consequence of Theorem 1.18.

**Theorem 1.20.** *If $f \colon D \to \mathbb{R}^n$ is a function of class $C^1$ in an open set $D \subset \mathbb{R} \times \mathbb{R}^n$, then for each $(t_0, x_0) \in D$ there exists a unique solution of the initial value problem* (1.13) *in some open interval containing $t_0$.*

**Proof.** The result follows immediately from Theorem 1.18 together with Proposition 1.19. $\square$

However, locally Lipschitz functions need not be differentiable.

**Example 1.21.** For each $x, y \in \mathbb{R}$, we have

$$\big| |x| - |y| \big| \leq |x - y|.$$

Hence, the function $f \colon \mathbb{R}^2 \to \mathbb{R}$ defined by $f(t, x) = |x|$ is locally Lipschitz in $x$, taking $L = 1$ in (1.25) for every compact set $K \subset \mathbb{R}^2$.

We also give an example of a function that is not locally Lipschitz.

**Example 1.22.** For the function $f \colon \mathbb{R}^2 \to \mathbb{R}$ defined by $f(t, x) = x^{2/3}$, we have

$$|f(t, x) - f(t, 0)| = |x^{2/3} - 0^{2/3}| = \frac{1}{|x|^{1/3}} |x - 0|.$$

Since $1/|x|^{1/3} \to +\infty$ when $x \to 0$, we conclude that $f$ is not locally Lipschitz in $x$ in any open set $D \subset \mathbb{R}^2$ intersecting the line $\mathbb{R} \times \{0\}$.

The following example illustrates that if $f$ is continuous but not locally Lipschitz, then the initial value problem (1.13) may not have a unique solution.

**Example 1.23** (continuation of Example 1.22). If $x = x(t)$ is a solution of the equation $x' = x^{2/3}$, then in any open interval where the solution does not take the value zero, we have $x'/x^{2/3} = 1$. Integrating on both sides, we obtain $x(t) = (t + c)^3/27$ for some constant $c \in \mathbb{R}$. One can easily verify by direct substitution in the equation that each of these solutions is defined in $\mathbb{R}$ (even if vanishing at $t = -c$). In particular, $x(t) = t^3/27$ is a solution of the initial value problem

$$\begin{cases} x' = x^{2/3}, \\ x(0) = 0, \end{cases}$$

which also has the constant solution $x(t) = 0$.

On the other hand, we show in Section 1.4 that for a continuous function $f$ the initial value problem (1.13) always has a solution, although it need not be unique.

**1.2.2. Contractions in metric spaces.** This section contains material of a more transverse nature, and is primarily included for completeness of the exposition. In particular, the theory is developed in a pragmatic manner having in mind the theory of differential equations.

We first recall some basic notions.

**Definition 1.24.** Given a set $X$, a function $d \colon X \times X \to \mathbb{R}_0^+$ is said to be a *distance* in $X$ if:

  a) $d(x, y) = 0$ if and only if $x = y$;
  b) $d(x, y) = d(y, x)$ for every $x, y \in X$;
  c) $d(x, y) \le d(x, z) + d(z, y)$ for every $x, y, z \in X$.

We then say that the pair $(X, d)$ is a *metric space*.

For example, one can define a distance in $\mathbb{R}^n$ by

$$d\big((x_1, \ldots, x_n), (y_1, \ldots, y_n)\big) = \left( \sum_{i=1}^{n} (x_i - y_i)^2 \right)^{1/2}, \qquad (1.27)$$

or

$$d\big((x_1,\ldots,x_n),(y_1,\ldots,y_n)\big) = \max\big\{|x_i - y_i| : i = 1,\ldots,n\big\}, \qquad (1.28)$$

among many other possibilities.

**Definition 1.25.** Let $(X,d)$ be a metric space. Given $x \in X$ and $r > 0$, we define the *open ball* of radius $r$ centered at $x$ by

$$B(x,r) = \big\{y \in X : d(y,x) < r\big\},$$

and the *closed ball* of radius $r$ centered at $x$ by

$$\overline{B(x,r)} = \big\{y \in X : d(y,x) \le r\big\}.$$

A particular class of distances is obtained from the notion of norm.

**Definition 1.26.** Let $X$ be a linear space (over $\mathbb{R}$). A function $\|\cdot\| \colon X \to \mathbb{R}_0^+$ is said to be a *norm* in $X$ if:

a) $\|x\| = 0$ if and only if $x = 0$;

b) $\|\lambda x\| = |\lambda| \cdot \|x\|$ for every $\lambda \in \mathbb{R}$ and $x \in X$;

c) $\|x + y\| \le \|x\| + \|y\|$ for every $x,y \in X$.

We then say that the pair $(X, \|\cdot\|)$ is a *normed space*.

For example, one can define a norm in $\mathbb{R}^n$ by

$$\|(x_1,\ldots,x_n)\| = \left(\sum_{i=1}^n x_i^2\right)^{1/2} \qquad (1.29)$$

or

$$\|(x_1,\ldots,x_n)\| = \max\big\{|x_i| : i = 1,\ldots,n\big\}, \qquad (1.30)$$

among many other possibilities.

**Proposition 1.27.** *If the pair $(X, \|\cdot\|)$ is a normed space, then the function $d \colon X \times X \to \mathbb{R}_0^+$ defined by $d(x,y) = \|x - y\|$ is a distance in $X$.*

**Proof.** For the first property in the notion of distance, it is sufficient to observe that

$$d(x,y) = 0 \quad \Leftrightarrow \quad \|x - y\| = 0 \quad \Leftrightarrow \quad x - y = 0,$$

using the first property in the notion of norm. Hence, $d(x,y) = 0$ if and only if $x = y$. For the second property, we note that

$$\begin{aligned} d(y,x) &= \|y - x\| = \|-(x - y)\| \\ &= |-1| \cdot \|x - y\| = \|x - y\|, \end{aligned}$$

using the second property in the notion of norm. Hence, $d(y, x) = d(x, y)$.
Finally, for the last property, we observe that

$$\begin{aligned} d(x, y) = \|x - y\| &= \|(x - z) + (z - y)\| \\ &\leq \|x - z\| + \|z - y\| \\ &= d(x, z) + d(z, y), \end{aligned}$$

using the third property in the notion of norm.                              □

For example, the distances in (1.27) and (1.28) are obtained as described
in Proposition 1.27, respectively, from the norms in (1.29) and (1.30).

Now we introduce the notions of a convergent sequence and of a Cauchy
sequence.

**Definition 1.28.** Let $(x_n)_n = (x_n)_{n \in \mathbb{N}}$ be a sequence in a metric space
$(X, d)$. We say that:

a) $(x_n)_n$ is a *convergent sequence* if there exists $x \in X$ such that
$$d(x_n, x) \to 0 \quad \text{when} \quad n \to \infty;$$

b) $(x_n)_n$ is a *Cauchy sequence* if given $\varepsilon > 0$, there exists $p \in \mathbb{N}$ such
that
$$d(x_n, x_m) < \varepsilon \quad \text{for every} \quad n, m > p.$$

Clearly, any convergent sequence is a Cauchy sequence.

**Definition 1.29.** A metric space $(X, d)$ is said to be *complete* if any Cauchy
sequence in $X$ is convergent.

For example, the space $\mathbb{R}^n$ with the distance in (1.27) or the distance
in (1.28) is a complete metric space. The following is another example.

**Proposition 1.30.** *The set $X = C(I)$ of all bounded continuous functions*
$x \colon I \to \mathbb{R}^n$ *in a set $I \subset \mathbb{R}^k$ is a complete metric space with the distance*

$$d(x, y) = \sup \left\{ \|x(t) - y(t)\| : t \in I \right\}. \tag{1.31}$$

**Proof.** One can easily verify that $d$ is a distance. In order to show that $X$
is a complete metric space, let $(x_p)_p$ be a Cauchy sequence in $X$. For each
$t \in I$, we have

$$\|x_p(t) - x_q(t)\| \leq d(x_p, x_q), \tag{1.32}$$

and thus $(x_p(t))_p$ is a Cauchy sequence in $\mathbb{R}^n$. Hence, it is convergent (because all Cauchy sequences in $\mathbb{R}^n$ are convergent) and there exists the limit

$$x(t) = \lim_{p \to \infty} x_p(t). \tag{1.33}$$

This yields a function $x \colon I \to \mathbb{R}^n$. For each $t, s \in I$, we have

$$\|x(t) - x(s)\| \leq \|x(t) - x_p(t)\| + \|x_p(t) - x_p(s)\| + \|x_p(s) - x(s)\|. \tag{1.34}$$

On the other hand, it follows from (1.32) (and the fact that $(x_p)_p$ is a Cauchy sequence) that given $\varepsilon > 0$, there exists $r \in \mathbb{N}$ such that

$$\|x_p(t) - x_q(t)\| < \varepsilon \qquad (1.35)$$

for every $t \in I$ and $p, q \geq r$. Letting $q \to \infty$ in (1.35), we obtain

$$\|x_p(t) - x(t)\| \leq \varepsilon \qquad (1.36)$$

for every $t \in I$ and $p \geq r$. Hence, it follows from (1.34) (taking $p = r$) that

$$\|x(t) - x(s)\| \leq 2\varepsilon + \|x_r(t) - x_r(s)\| \qquad (1.37)$$

for every $t, s \in I$. Since the function $x_r$ is continuous, given $t \in I$, there exists $\delta > 0$ such that

$$\|x_r(t) - x_r(s)\| < \varepsilon \quad \text{whenever} \quad \|t - s\| < \delta.$$

Hence, it follows from (1.37) that

$$\|x(t) - x(s)\| < 3\varepsilon \quad \text{whenever} \quad \|t - s\| < \delta,$$

and the function $x$ is continuous. Moreover, it follows from (1.36) that given $\varepsilon > 0$, there exists $r \in \mathbb{N}$ such that

$$\|x(t)\| \leq \|x_p(t) - x(t)\| + \|x_p(t)\|$$
$$\leq \varepsilon + \sup\left\{\|x_p(t)\| : t \in I\right\} < +\infty$$

for every $p \geq r$, and hence $x \in X$. Furthermore, also by (1.36),

$$d(x_p, x) = \sup\left\{\|x_p(t) - x(t)\| : t \in I\right\} \leq \varepsilon$$

for every $p \geq r$, and thus $d(x_p, x) \to 0$ when $p \to \infty$. This shows that $X$ is a complete metric space. $\qquad \square$

Now we recall the notion of Lipschitz function.

**Definition 1.31.** A function $x\colon I \to \mathbb{R}^n$ in a set $I \subset \mathbb{R}^k$ is said to be *Lipschitz* if there exists $L > 0$ such that

$$\|x(t) - x(s)\| \leq L\|t - s\| \qquad (1.38)$$

for every $t, s \in I$.

Clearly, all Lipschitz functions are continuous.

**Proposition 1.32.** *The set $Y \subset C(I)$ of all bounded Lipschitz functions with the same constant $L$ in (1.38) is a complete metric space with the distance $d$ in (1.31).*

**Proof.** In view of Proposition 1.30, it is sufficient to show that if $(x_p)_p$ is a Cauchy sequence in $Y$ (and thus also in $C(I)$), then its limit $x$ satisfies (1.38). Then, let $(x_p)_p$ be a Cauchy sequence in $Y$. We have

$$\|x_p(t) - x_p(s)\| \le L\|t - s\|, \quad t, s \in I \tag{1.39}$$

for every $p \in \mathbb{N}$. On the other hand, by (1.33), for each $t \in I$ we have $x_p(t) \to x(t)$ when $p \to \infty$, and thus it follows from (1.39) that $x \in Y$. $\square$

Now we consider a particular class of transformations occurring several times in the proofs of this book.

**Definition 1.33.** A transformation $T\colon X \to X$ in a metric space $(X, d)$ is said to be a *contraction* if there exists $\lambda \in (0, 1)$ such that

$$d(T(x), T(y)) \le \lambda\, d(x, y)$$

for every $x, y \in X$.

**Example 1.34.** Let $A$ be an $n \times n$ matrix and let $T\colon \mathbb{R}^n \to \mathbb{R}^n$ be the linear transformation defined by $T(x) = Ax$. Consider the distance $d$ in (1.27), obtained from the norm

$$\|(x_1, \ldots, x_n)\| = \left( \sum_{i=1}^{n} x_i^2 \right)^{1/2}.$$

We have

$$d(T(x), T(y)) = \|Ax - Ay\|$$
$$= \|A(x - y)\| \le \|A\| d(x, y),$$

where

$$\|A\| = \sup_{x \ne 0} \frac{\|Ax\|}{\|x\|}. \tag{1.40}$$

In particular, if $\|A\| < 1$, then the transformation $T$ is a contraction.

For example, if $A$ is a diagonal matrix with entries $\lambda_1, \ldots, \lambda_n$ in the diagonal, then

$$\|A\| = \sup_{(x_1, \ldots, x_n) \ne 0} \frac{\|(\lambda_1 x_1, \ldots, \lambda_n x_n)\|}{\|(x_1, \ldots, x_n)\|}$$
$$\le \max\big\{|\lambda_i| : i = 1, \ldots, n\big\}.$$

Thus, if $|\lambda_i| < 1$ for $i = 1, \ldots, n$, then $\|A\| < 1$ and $T$ is a contraction.

We say that $x_0 \in X$ is a *fixed point* of a transformation $T\colon X \to X$ if $T(x_0) = x_0$. Contractions satisfy the following fixed point theorem, where $T^n$ is defined recursively by $T^{n+1} = T \circ T^n$ for $n \in \mathbb{N}$.

**Theorem 1.35.** *If $T\colon X \to X$ is a contraction in a complete metric space $(X, d)$, then $T$ has a unique fixed point. Moreover, for each $x \in X$ the sequence $(T^n(x))_n$ converges to the unique fixed point of $T$.*

**Proof.** Given $x \in X$, consider the sequence $x_n = T^n(x)$ for $n \in \mathbb{N}$. For each $m, n \in \mathbb{N}$ with $m > n$, we have

$$
\begin{aligned}
d(x_m, x_n) &\le d(x_m, x_{m-1}) + d(x_{m-1}, x_{m-2}) + \cdots + d(x_{n+1}, x_n) \\
&\le (\lambda^{m-1} + \lambda^{m-2} + \cdots + \lambda^n) d(T(x), x) \\
&= \lambda^n \frac{1 - \lambda^{m-n}}{1 - \lambda} d(T(x), x) \\
&\le \frac{\lambda^n}{1 - \lambda} d(T(x), x).
\end{aligned}
\tag{1.41}
$$

Thus, $(x_n)_n$ is a Cauchy sequence in $X$, and since the space is complete the sequence has a limit, say $x_0 \in X$. Now we note that

$$
d(T(x_n), T(x_0)) \le \lambda\, d(x_n, x_0) \to 0
$$

when $n \to \infty$, and hence, the sequence $(T(x_n))_n$ converges to $T(x_0)$. But since $T(x_n) = x_{n+1}$, the sequence $(T(x_n))_n$ also converges to $x_0$. It follows from the uniqueness of the limit that $T(x_0) = x_0$, that is, $x_0$ is a fixed point of the transformation $T$.

In order to show that the fixed point is unique, let us assume that $y_0 \in X$ is also a fixed point. Then

$$
d(x_0, y_0) = d(T(x_0), T(y_0)) \le \lambda\, d(x_0, y_0).
\tag{1.42}
$$

Since $\lambda < 1$, it follows from (1.42) that $x_0 = y_0$. The last property in the theorem follows from the uniqueness of the fixed point, since we have already shown that for each $x \in X$ the sequence $(T^n(x))_n$ converges to a fixed point of $T$. $\qquad\square$

It follows from the proof of Theorem 1.35 (see (1.41)) that if $x_0 \in X$ is the unique fixed point of a contraction $T\colon X \to X$, then

$$
d(T^n(x), x_0) \le \frac{\lambda^n}{1 - \lambda} d(T(x), x)
$$

for every $x \in X$ and $n \in \mathbb{N}$. In particular, each sequence $(T^n(x))_n$ converges exponentially to $x_0$.

Motivated by the last property in Theorem 1.35, we introduce the following notion.

**Definition 1.36.** A fixed point $x_0 \in X$ of a transformation $T\colon X \to X$ is said to be *attracting* if for each $x \in X$ we have $T^n(x) \to x_0$ when $n \to \infty$.

Now we consider a more general situation in which it is still possible to establish a fixed point theorem. Given metric spaces $X$ and $Y$ (not necessarily complete), consider a transformation $S\colon X \times Y \to X \times Y$ of the form

$$S(x,y) = (T(x), A(x,y)), \tag{1.43}$$

where $T\colon X \to X$ and $A\colon X \times Y \to Y$. Each set $\{x\} \times Y$ is called a *fiber* of $X \times Y$. We note that the image of the fiber $\{x\} \times Y$ under the transformation $S$ is contained in the fiber $\{T(x)\} \times Y$.

**Definition 1.37.** The transformation $S$ in (1.43) is said to be a *fiber contraction* if there exists $\lambda \in (0,1)$ such that

$$d_Y(A(x,y), A(x,\bar{y})) \le \lambda\, d_Y(y,\bar{y}) \tag{1.44}$$

for every $x \in X$ and $y, \bar{y} \in Y$.

The following is a fixed point theorem for fiber contractions.

**Theorem 1.38** (Fiber contraction theorem). *For a continuous fiber contraction $S$, if $x_0 \in X$ is an attracting fixed point of $T$ and $y_0 \in Y$ is a fixed point of $y \mapsto A(x_0, y)$, then $(x_0, y_0)$ is an attracting fixed point of $S$.*

**Proof.** Let $d_X$ and $d_Y$ be, respectively, the distances in $X$ and $Y$. We define a distance $d$ in $X \times Y$ by

$$d\big((x,y), (\bar{x}, \bar{y})\big) = d_X(x, \bar{x}) + d_Y(y, \bar{y}).$$

It follows from the triangle inequality that

$$d\big(S^n(x,y), (x_0, y_0)\big) \le d\big(S^n(x,y), S^n(x, y_0)\big) + d\big(S^n(x, y_0), (x_0, y_0)\big). \tag{1.45}$$

We want to show that both terms in the right-hand side of (1.45) tend to zero when $n \to \infty$. Given $x \in X$, define a transformation $A_x\colon Y \to Y$ by $A_x(y) = A(x,y)$. Clearly,

$$S(x,y) = \big(T(x), A_x(y)\big).$$

Moreover, for each $n \in \mathbb{N}$ we have

$$S^n(x,y) = \big(T^n(x), A_{x,n}(y)\big), \tag{1.46}$$

where

$$A_{x,n} = A_{T^{n-1}(x)} \circ \cdots \circ A_{T(x)} \circ A_x.$$

For the first term in (1.45), it follows from (1.44) and (1.46) that

$$d\big(S^n(x,y), S^n(x, y_0)\big) = d_Y\big(A_{x,n}(y), A_{x,n}(y_0)\big) \le \lambda^n d_Y(y, y_0) \to 0 \tag{1.47}$$

when $n \to \infty$. For the second term, we note that

$$d\big(S^n(x, y_0), (x_0, y_0)\big) \le d_X(T^n(x), x_0) + d_Y\big(A_{x,n}(y_0), y_0\big). \tag{1.48}$$

Since $x_0$ is an attracting fixed point of $T$, we have $d_X(T^n(x), x_0) \to 0$ when $n \to \infty$. In order to show that $(x_0, y_0)$ is an attracting fixed point of $S$, it remains to verify that the second term in the right-hand side of (1.48) tends to zero when $n \to \infty$. To that effect, note that

$$d_Y(A_{x,n}(y_0), y_0)$$

$$\leq \sum_{i=0}^{n-1} d_Y\big((A_{T^{n-1}(x)} \circ \cdots \circ A_{T^i(x)})(y_0), (A_{T^{n-1}(x)} \circ \cdots \circ A_{T^{i+1}(x)})(y_0)\big)$$

$$\leq \sum_{i=0}^{n-1} \lambda^{n-i-1} d_Y(A_{T^i(x)}(y_0), y_0),$$

where $A_{T^{n-1}(x)} \circ \cdots \circ A_{T^{i+1}(x)} = \mathrm{Id}$ for $i = n-1$. Since $y_0$ is a fixed point of $A_{x_0}$, we have

$$c_i := d_Y\big(A_{T^i(x)}(y_0), y_0\big) = d_Y\big(A(T^i(x), y_0), A(x_0, y_0)\big). \tag{1.49}$$

On the other hand, since the transformation $A$ is continuous (because $S$ is continuous) and $x_0$ is an attracting fixed point of $T$, it follows from (1.49) that the sequence $c_i$ converges to zero when $i \to \infty$. In particular, there exists $c > 0$ such that $0 \leq c_i < c$ for every $i \in \mathbb{N}$. Thus, given $k \in \mathbb{N}$ and $n \geq k$, we have

$$\sum_{i=0}^{n} \lambda^{n-i} c_i = \sum_{i=0}^{k-1} \lambda^{n-i} c_i + \sum_{i=k}^{n} \lambda^{n-i} c_i$$

$$\leq c \sum_{i=0}^{k-1} \lambda^{n-i} + \sup_{j \geq k} c_j \sum_{i=k}^{n} \lambda^{n-i}$$

$$\leq c \frac{\lambda^{n-k+1}}{1-\lambda} + \sup_{j \geq k} c_j \frac{1}{1-\lambda}.$$

Therefore,

$$\limsup_{n \to \infty} \sum_{i=0}^{n} \lambda^{n-i} c_i \leq \sup_{j \geq k} c_j \frac{1}{1-\lambda} \to 0$$

when $k \to \infty$. This shows that

$$d_Y(A_{x,n}(y_0), y_0) \to 0 \quad \text{when} \quad n \to \infty,$$

and it follows from (1.45), (1.47) and (1.48) that $(x_0, y_0)$ is an attracting fixed point of $S$. □

**1.2.3. Proof of the theorem.** Now we use the theory developed in the previous section to prove the Picard–Lindelöf theorem.

**Proof of Theorem 1.18.** By Proposition 1.9, the initial value problem (1.13) consists of finding a function $x \in C(a, b)$ in an open interval $(a, b)$

containing $t_0$ such that

$$x(t) = x_0 + \int_{t_0}^{t} f(s, x(s)) \, ds \qquad (1.50)$$

for every $t \in (a, b)$. Here $C(a, b)$ is the set of all bounded continuous functions $y \colon (a, b) \to \mathbb{R}^n$. We look for $x$ as a fixed point of a contraction.

Take constants $a < t_0 < b$ and $\beta > 0$ such that

$$K := [a, b] \times \overline{B(x_0, \beta)} \subset D, \qquad (1.51)$$

where

$$\overline{B(x_0, \beta)} = \{y \in \mathbb{R}^n : \|y - x_0\| \le \beta\}.$$

Also, let $X \subset C(a, b)$ be the set of all continuous functions $x \colon (a, b) \to \mathbb{R}^n$ such that

$$\|x(t) - x_0\| \le \beta \quad \text{for} \quad t \in (a, b).$$

We first show that $X$ is a complete metric space with the distance in (1.31). Given a Cauchy sequence $(x_p)_p$ in $X$, it follows from Proposition 1.30 that it converges to a function $x \in C(a, b)$. In order to show that $x \in X$, we note that

$$\|x(t) - x_0\| = \lim_{p \to \infty} \|x_p(t) - x_0\| \le \beta$$

for $t \in (a, b)$, since $\|x_p(t) - x_0\| \le \beta$ for $t \in (a, b)$ and $p \in \mathbb{N}$. Moreover, there is no loss of generality in looking for fixed points in $X$ and not in $C(a, b)$. Indeed, if $x \colon (a, b) \to \mathbb{R}^n$ is a continuous function satisfying (1.50), then

$$\|x(t) - x_0\| \le \left\| \int_{t_0}^{t} f(s, x(s)) \, ds \right\|$$
$$\le |t - t_0| M \le (b - a) M,$$

where

$$M = \max \{\|f(t, x)\| : (t, x) \in K\} < +\infty, \qquad (1.52)$$

because $f$ is continuous and $K$ is compact (recall that any continuous function with values in $\mathbb{R}$, such as $(t, x) \mapsto \|f(t, x)\|$, has a maximum in each compact set). This shows that if $x \in C(a, b)$ satisfies (1.50), then it belongs to $X$ for some $\beta$.

Now we consider the transformation $T$ defined by

$$T(x)(t) = x_0 + \int_{t_0}^{t} f(s, x(s)) \, ds$$

for each $x \in X$. We note that $t \mapsto T(x)(t)$ is continuous and

$$\|T(x)(t) - x_0\| \le \left\| \int_{t_0}^{t} f(s, x(s)) \, ds \right\| \le (b - a) M.$$

For $b - a$ sufficiently small, we have $(b-a)M \leq \beta$ and thus $T(X) \subset X$. Moreover, given $x, y \in X$,

$$
\begin{aligned}
\|T(x)(t) - T(y)(t)\| &\leq \left\| \int_{t_0}^t [f(s, x(s)) - f(s, y(s))] \, ds \right\| \\
&\leq \left| \int_{t_0}^t L \|x(s) - y(s)\| \, ds \right| \\
&\leq (b-a) L \, d(x, y),
\end{aligned}
$$

where $L$ is the constant in (1.25) for the compact set $K$ in (1.51) and $d$ is the distance in (1.31) for $I = (a, b)$. Hence,

$$
d(T(x), T(y)) \leq (b-a) L \, d(x, y)
$$

for every $x, y \in X$. For $b - a$ sufficiently small, we have $(b-a)L < 1$, in addition to $(b-a)M \leq \beta$, and $T$ is a contraction in the complete metric space $X$. By Theorem 1.35, we conclude that $T$ has a unique fixed point $x \in X$. This is the unique continuous function in $(a, b)$ satisfying (1.50). $\quad \square$

## 1.3. Additional properties

In this section we study some additional properties of the solutions of an ordinary differential equation, including the dependence of the solutions on the initial conditions and what happens when a given solution cannot be extended to a larger interval.

### 1.3.1. Lipschitz dependence on the initial conditions. We first describe how the solutions depend on the initial conditions for a continuous function $f$ that is locally Lipschitz.

We start with an auxiliary result.

**Proposition 1.39** (Gronwall's lemma). *Let $u, v \colon [a, b] \to \mathbb{R}^n$ be continuous functions with $v \geq 0$ and let $c \in \mathbb{R}$. If*

$$
u(t) \leq c + \int_a^t u(s) v(s) \, ds \tag{1.53}
$$

*for every $t \in [a, b]$, then*

$$
u(t) \leq c \exp \int_a^t v(s) \, ds
$$

*for every $t \in [a, b]$.*

**Proof.** Consider the functions

$$
R(t) = \int_a^t u(s) v(s) \, ds
$$

and

$$V(t) = \int_a^t v(s)\, ds.$$

Clearly, $R(a) = 0$ and by hypothesis,

$$R'(t) = u(t)v(t) \le (c + R(t))v(t).$$

Therefore,

$$R'(t) - v(t)R(t) \le cv(t)$$

and

$$\frac{d}{dt}\left(e^{-V(t)}R(t)\right) = e^{-V(t)}\left(R'(t) - v(t)R(t)\right)$$

$$\le cv(t)e^{-V(t)}.$$

Since $R(a) = 0$, we obtain

$$e^{-V(t)}R(t) \le \int_a^t cv(\tau)e^{-V(\tau)}\, d\tau$$

$$= -ce^{-V(\tau)}\Big|_{\tau=a}^{\tau=t} = c\left(1 - e^{-V(t)}\right),$$

and hence,

$$R(t) \le ce^{V(t)} - c.$$

Thus, it follows from (1.53) that

$$u(t) \le c + R(t) \le ce^{V(t)}$$

for every $t \in [a, b]$. This yields the desired result.                    $\square$

Now we establish the Lipschitz dependence of the solutions on the initial conditions.

**Theorem 1.40.** *Let $f\colon D \to \mathbb{R}^n$ be continuous and locally Lipschitz in $x$ in an open set $D \subset \mathbb{R} \times \mathbb{R}^n$. Given $(t_0, x_1) \in D$, there exist constants $\beta, C > 0$ and an open interval $I$ containing $t_0$ such that for each $x_2 \in \mathbb{R}^n$ with $\|x_1 - x_2\| < \beta$ the solutions $x_1(t)$ and $x_2(t)$, respectively, of the initial value problems*

$$\begin{cases} x' = f(t, x), \\ x(t_0) = x_1 \end{cases} \quad \text{and} \quad \begin{cases} x' = f(t, x), \\ x(t_0) = x_2 \end{cases} \tag{1.54}$$

*satisfy*

$$\|x_1(t) - x_2(t)\| \le C\|x_1 - x_2\| \quad \text{for} \quad t \in I. \tag{1.55}$$

**Proof.** By the Picard–Lindelöf theorem (Theorem 1.18), the initial value problems in (1.54) have unique solutions $x_1(t)$ and $x_2(t)$ in some open interval $I$ containing $t_0$ (which we can assume to be the same). Moreover, by Proposition 1.9, we have

$$x_i(t) = x_i + \int_{t_0}^{t} f(s, x(s)) \, ds \qquad (1.56)$$

for $t \in I$ and $i = 1, 2$. For simplicity of the exposition, we consider only times $t \geq t_0$. The case when $t \leq t_0$ can be treated in an analogous manner.

We define a function $y \colon I \to \mathbb{R}^n$ by

$$y(t) = x_1(t) - x_2(t).$$

For $t \geq t_0$, it follows from (1.56) that

$$\|y(t)\| \leq \|x_1 - x_2\| + \int_{t_0}^{t} \|f(s, x_1(s)) - f(s, x_2(s))\| \, ds$$

$$\leq \|x_1 - x_2\| + L \int_{t_0}^{t} \|y(s)\| \, ds,$$

for some constant $L$ (because $f$ is locally Lipschitz in $x$). Letting

$$u(t) = \|y(t)\|, \quad v(t) = L \quad \text{and} \quad c = \|x_1 - x_2\|,$$

it follows from Gronwall's lemma (Proposition 1.39) that

$$u(t) \leq c \exp \int_{t_0}^{t} v(s) \, ds = \|x_1 - x_2\| e^{L(t - t_0)}$$

for $t \in I \cap [t_0, +\infty)$, which establishes the inequality in (1.55) for these values of $t$. $\qquad \square$

**1.3.2. Smooth dependence on the initial conditions.** In this section we show that for a function $f$ of class $C^1$ the solutions are also of class $C^1$ on the initial conditions. The proof uses the Fiber contraction theorem (Theorem 1.38).

We first establish an auxiliary result on the limit of a sequence of differentiable functions.

**Proposition 1.41.** *If a sequence of differentiable functions $f_n \colon U \to \mathbb{R}^n$ in an open set $U \subset \mathbb{R}^n$ is (pointwise) convergent and the sequence $df_n$ of their derivatives is uniformly convergent, then the limit of the sequence $f_n$ is differentiable in $U$ and its derivative is the limit of the sequence $df_n$ in $U$.*

**Proof.** Let $f \colon U \to \mathbb{R}^n$ and $g \colon U \to M_n$ be, respectively, the (pointwise) limits of the sequences $f_n$ and $df_n$, where $M_n$ is the set of $n \times n$ matrices with real entries. Given $x, h \in \mathbb{R}^n$ such that the line segment between $x$ and $x + h$ is contained in $U$, we have

$$f_n(x + h) - f_n(x) = \int_0^1 \frac{d}{dt} f_n(x + th) \, dt = \int_0^1 d_{x+th} f_n h \, dt \qquad (1.57)$$

for every $n \in \mathbb{N}$. On the other hand, since the sequence $df_n$ converges uniformly, given $\delta > 0$, there exists $p \in \mathbb{N}$ such that

$$\sup \{ \| d_y f_n - d_y f_m \| : y \in U \} < \delta$$

for $m, n > p$. Hence, it follows from (1.57) that

$$\begin{aligned}
&\left\| \left( f_n(x + h) - f_n(x) \right) - \left( f_m(x + h) - f_m(x) \right) \right\| \\
&= \left\| \int_0^1 \left( d_{x+th} f_n - d_{x+th} f_m \right) h \, dt \right\| \\
&\leq \int_0^1 \| d_{x+th} f_n - d_{x+th} f_m \| \, dt \| h \| \leq \delta \| h \|
\end{aligned}$$

for $m, n > p$. Letting $m \to \infty$, we obtain

$$\left\| \left( f_n(x + h) - f_n(x) \right) - \left( f(x + h) - f(x) \right) \right\| \leq \delta \| h \| \qquad (1.58)$$

for every $n > p$.

Now we show that $f$ is differentiable. Given $x, h \in \mathbb{R}^n$ such that the line segment between $x$ and $x + h$ is contained in $U$, for each $\varepsilon \in (0, 1)$ we have

$$\begin{aligned}
\left\| \frac{f(x + \varepsilon h) - f(x)}{\varepsilon} - g(x) h \right\| &\leq \left\| \frac{f(x + \varepsilon h) - f(x)}{\varepsilon} - \frac{f_n(x + \varepsilon h) - f_n(x)}{\varepsilon} \right\| \\
&+ \left\| \frac{f_n(x + \varepsilon h) - f_n(x)}{\varepsilon} - d_x f_n h \right\| \\
&+ \| d_x f_n h - g(x) h \|.
\end{aligned}$$
$$(1.59)$$

Since $g$ is the limit of the sequence $df_n$, given $\delta > 0$, there exists $q \in \mathbb{N}$ such that

$$\sup \{ \| d_y f_n - g(y) \| : y \in U \} < \delta \qquad (1.60)$$

for $n > q$. Now, take $n > \max\{p, q\}$. Since $f_n$ is differentiable, for any sufficiently small $\varepsilon$ we have

$$\left\| \frac{f_n(x + \varepsilon h) - f_n(x)}{\varepsilon} - d_x f_n h \right\| < \delta \| h \|. \qquad (1.61)$$

It follows from (1.59) together with (1.58), (1.60) and (1.61) that

$$\left\| \frac{f(x + \varepsilon h) - f(x)}{\varepsilon} - g(x) h \right\| \leq 3 \delta \| h \|.$$

Therefore, the function $f$ is differentiable at $x$, and $d_x f h = g(x) h$ for each $h \in \mathbb{R}^n$, that is, $df = g$. $\qquad \square$

Now we establish the $C^1$ dependence of the solutions on the initial conditions.

**Theorem 1.42.** *If the function $f: D \to \mathbb{R}^n$ is of class $C^1$ in an open set $D \subset \mathbb{R} \times \mathbb{R}^n$, then for each $(t_0, x_0) \in D$ the function $(t, x) \mapsto \varphi(t, x)$, where $\varphi(\cdot, x_0)$ is the solution of the initial value problem (1.13), is of class $C^1$ in a neighborhood of $(t_0, x_0)$.*

**Proof.** We divide the proof into several steps.

*Step 1. Construction of a metric space.* Given $(t_0, x_0) \in D$, take constants $\alpha, \beta > 0$ such that

$$S := [t_0 - \alpha, t_0 + \alpha] \times \overline{B(x_0, 2\beta)} \subset D. \tag{1.62}$$

Since the set $S$ is compact and $f$ is continuous, we have

$$K := \max \left\{ \|f(t, x)\| : (t, x) \in S \right\} < +\infty$$

(recall that a continuous function with values in $\mathbb{R}$ has a maximum in each compact set). If necessary, we take a smaller $\alpha > 0$ such that $\alpha K \le \beta$, and we consider the set $X$ of all continuous functions $\varphi: U \to B(x_0, 2\beta)$ in the open set

$$U = (t_0 - \alpha, t_0 + \alpha) \times B(x_0, \beta), \tag{1.63}$$

equipped with the distance

$$d(\varphi, \psi) = \sup \left\{ \|\varphi(t, x) - \psi(t, x)\| : (t, x) \in U \right\}. \tag{1.64}$$

Repeating arguments in the proof of the Picard–Lindelöf theorem (Theorem 1.18), it follows easily from Proposition 1.30 that $X$ is a complete metric space.

*Step 2. Construction of a contraction in $X$.* We define a transformation $T$ in $X$ by

$$T(\varphi)(t, x) = x + \int_{t_0}^{t} f(s, \varphi(s, x)) \, ds$$

for each $(t, x) \in U$. We first show that $T(\varphi)$ is a continuous function. Clearly, $t \mapsto T(\varphi)(t, x)$ is continuous for each $x \in B(x_0, \beta)$. Given $t \in (t_0 - \alpha, t_0 + \alpha)$, we show that the function $x \mapsto T(\varphi)(t, x)$ is also continuous. For each $x, y \in B(x_0, \beta)$, we have

$$\|T(\varphi)(t, x) - T(\varphi)(t, y)\| \le \|x - y\| + \left| \int_{t_0}^{t} \|f(s, \varphi(s, x)) - f(s, \varphi(s, y))\| \, ds \right|$$

$$\le \|x - y\| + L \left| \int_{t_0}^{t} \|\varphi(s, x) - \varphi(s, y)\| \, ds \right| \tag{1.65}$$

for some constant $L$. Indeed, by Proposition 1.19, the function $f$ is locally Lipschitz in $x$, and hence, one can take $L$ to be the constant in (1.25) for the compact set $S$ in (1.62). Moreover, each function $\varphi \in X$ is uniformly continuous on the compact set

$$\overline{U} = [t_0 - \alpha, t_0 + \alpha] \times \overline{B(x_0, \beta)},$$

because $\varphi$ is continuous. Thus, given $\delta > 0$, there exists $\varepsilon > 0$ such that

$$\|\varphi(s, x) - \varphi(s, y)\| < \delta$$

for every $s \in [t_0 - \alpha, t_0 + \alpha]$ and $\|x - y\| < \varepsilon$. Without loss of generality, one can always take $\varepsilon < \delta$. Hence, it follows from (1.65) that

$$\|T(\varphi)(t, x) - T(\varphi)(t, y)\| < \varepsilon + L|t - t_0|\delta < \delta(1 + L\alpha)$$

whenever $\|x - y\| < \varepsilon$. This shows that the function $x \mapsto T(\varphi)(t, x)$ is continuous. On the other hand,

$$\|T(\varphi)(t, x) - x_0\| \leq \|x - x_0\| + \left| \int_{t_0}^{t} \|f(s, \varphi(s, x))\| \, ds \right|$$

$$\leq \|x - x_0\| + \alpha K < \beta + \beta = 2\beta,$$

and hence $T(X) \subset X$.

Now we verify that $T$ is a contraction. We have

$$\|T(\varphi)(t, x) - T(\psi)(t, x)\| \leq \left| \int_{t_0}^{t} \|f(s, \varphi(s, x)) - f(s, \psi(s, x))\| \, ds \right| \tag{1.66}$$

$$\leq L\alpha \, d(\varphi, \psi),$$

with the same constant $L$ as in (1.65), and thus,

$$d(T(\varphi), T(\psi)) \leq L\alpha \, d(\varphi, \psi).$$

For any sufficiently small $\alpha$, we have $L\alpha < 1$, in addition to $\alpha K \leq \beta$, and thus $T$ is a contraction. By Theorem 1.35, we conclude that $T$ has a unique fixed point $\varphi_0 \in X$, which thus satisfies the identity

$$\varphi_0(t, x) = x + \int_{t_0}^{t} f(s, \varphi_0(s, x)) \, ds \tag{1.67}$$

for every $t \in (t_0 - \alpha, t_0 + \alpha)$ and $x \in B(x_0, \beta)$. By Proposition 1.9, it follows from (1.67) that the function $t \mapsto \varphi_0(t, x_0)$ is a solution of the initial value problem (1.13). In particular it is of class $C^1$. To complete the proof, it remains to show that $\varphi_0$ is of class $C^1$ in $x$.

*Step 3. Construction of a fiber contraction.* Again, let $M_n$ be the set of $n \times n$ matrices with real entries. Also, let $Y$ be the set of all continuous functions $\Phi \colon U \to M_n$, with $U$ as in (1.63), such that

$$\sup \left\{ \|\Phi(t, x)\| : (t, x) \in U \right\} < +\infty$$

(the norm $\|\Phi(t,x)\|$ is defined by (1.40)). It follows from Proposition 1.30 that $Y$ is a complete metric space with the distance

$$d(\Phi, \Psi) := \sup \left\{ \|\Phi(t,x) - \Psi(t,x)\| : (t,x) \in U \right\}.$$

Now we define a transformation $A$ in $X \times Y$ by

$$A(\varphi, \Phi)(t,x) = \mathrm{Id} + \int_{t_0}^{t} \frac{\partial f}{\partial x}(s, \varphi(s,x))\Phi(s,x)\,ds,$$

where $\mathrm{Id} \in M_n$ is the identity matrix. We also consider the transformation $S$ in (1.43), that is,

$$S(\varphi, \Phi) = (T(\varphi), A(\varphi, \Phi)).$$

We first show that $S$ is continuous. By (1.66), the transformation $T$ is continuous. Moreover,

$$\|A(\varphi, \Phi)(t,x) - A(\psi, \Psi)(t,x)\|$$
$$= \left\| \int_{t_0}^{t} \frac{\partial f}{\partial x}(s, \varphi(s,x))\Phi(s,x)\,ds - \int_{t_0}^{t} \frac{\partial f}{\partial x}(s, \psi(s,x))\Psi(s,x)\,ds \right\|$$
$$\leq \left\| \int_{t_0}^{t} \frac{\partial f}{\partial x}(s, \varphi(s,x))[\Phi(s,x) - \Psi(s,x)]\,ds \right\| \qquad (1.68)$$
$$+ \left\| \int_{t_0}^{t} \left( \frac{\partial f}{\partial x}(s, \varphi(s,x)) - \frac{\partial f}{\partial x}(s, \psi(s,x)) \right) \Psi(s,x)\,ds \right\|$$
$$\leq \alpha M d(\Phi, \Psi)$$
$$+ \left| \int_{t_0}^{t} \left\| \frac{\partial f}{\partial x}(s, \varphi(s,x)) - \frac{\partial f}{\partial x}(s, \psi(s,x)) \right\| \cdot \|\Psi(s,x)\|\,ds \right|,$$

where

$$M := \max \left\{ \left\| \frac{\partial f}{\partial x}(t,x) \right\| : (t,x) \in \overline{U} \right\} < +\infty.$$

Taking $\varphi = \psi$ in (1.68), we obtain

$$d\big(A(\varphi, \Phi), A(\varphi, \Psi)\big) \leq \alpha M d(\Phi, \Psi). \qquad (1.69)$$

In particular, the function $\Phi \mapsto A(\varphi, \Phi)$ is continuous for each $\varphi \in X$. On the other hand, since $\partial f / \partial x$ is uniformly continuous on the compact set $\overline{U}$, given $\delta > 0$, there exists $\varepsilon > 0$ such that

$$\left\| \frac{\partial f}{\partial x}(s,x) - \frac{\partial f}{\partial x}(s,y) \right\| < \delta$$

for $s \in [t_0 - \alpha, t_0 + \alpha]$ and $\|x - y\| < \varepsilon$. Hence, if $d(\varphi, \psi) < \varepsilon$ (see (1.64)), then

$$\left\| \frac{\partial f}{\partial x}(s, \varphi(s,x)) - \frac{\partial f}{\partial x}(s, \psi(s,x)) \right\| < \delta$$

for $s \in (t_0 - \alpha, t_0 + \alpha)$ and $x \in B(x_0, \beta)$. It follows from (1.68) with $\Phi = \Psi$ that

$$\|A(\varphi, \Phi)(t, x) - A(\psi, \Phi)(t, x)\| \leq \alpha \delta \sup \{\|\Phi(t, x)\| : (t, x) \in U\},$$

and thus also

$$d\big(A(\varphi, \Phi), A(\psi, \Phi)\big) \leq \alpha \delta \sup \{\|\Phi(t, x)\| : (t, x) \in U\},$$

whenever $d(\varphi, \psi) < \varepsilon$. This shows that the function $\varphi \mapsto A(\varphi, \Phi)$ is continuous for each $\Phi \in Y$, and hence $A(X \times Y) \subset Y$. Moreover, for any sufficiently small $\alpha$ we have $\alpha M < 1$, and it follows from (1.69) that $S$ is a fiber contraction.

*Step 4. Uniform convergence and $C^1$ regularity.* By the Fiber contraction theorem (Theorem 1.38), given $(\varphi, \Phi) \in X \times Y$, the sequence $S^n(\varphi, \Phi)$ converges to $(\varphi_0, \Phi_0)$ when $n \to \infty$, where $\Phi_0$ is the unique fixed point of the contraction $\Phi \mapsto A(\varphi_0, \Phi)$ in the complete metric space $Y$. In other words, the sequence of functions $S^n(\varphi, \Phi)$ converges uniformly to the continuous function $(\varphi_0, \Phi_0)$ in the open set $U$ when $n \to \infty$.

Since $\Phi_0$ is continuous, if one can show that

$$\frac{\partial \varphi_0}{\partial x}(t, x) = \Phi_0(t, x), \tag{1.70}$$

then $\varphi_0$ is of class $C^1$ in $x$, and thus it is of class $C^1$. To that effect, consider the functions $\varphi_1$ and $\Phi_1$ given by $\varphi_1(t, x) = x$ and $\Phi_1(t, x) = \mathrm{Id}$. For each $n \in \mathbb{N}$, we define a pair of functions by

$$(\varphi_n, \Phi_n) = S^{n-1}(\varphi_1, \Phi_1).$$

These can be obtained recursively from the identities

$$\varphi_{n+1}(t, x) = x + \int_{t_0}^t f(s, \varphi_n(s, x)) \, ds \tag{1.71}$$

and

$$\Phi_{n+1}(t, x) = \mathrm{Id} + \int_{t_0}^t \frac{\partial f}{\partial x}(s, \varphi_n(s, x)) \Phi_n(s, x) \, ds. \tag{1.72}$$

One can use induction to show that the functions $\varphi_n$ are of class $C^1$ in $x$ for each $n \in \mathbb{N}$, with

$$
\begin{aligned}
\frac{\partial \varphi_{n+1}}{\partial x}(t, x) &= \frac{\partial}{\partial x}\left(x + \int_{t_0}^t f(s, \varphi_n(s, x)) \, ds\right) \\
&= \mathrm{Id} + \int_{t_0}^t \frac{\partial f}{\partial x}(s, \varphi_n(s, x)) \frac{\partial \varphi_n}{\partial x}(s, x) \, ds.
\end{aligned}
\tag{1.73}
$$

For this, it is sufficient to observe that if $\varphi_n$ is of class $C^1$ in $x$, then the function $(s, x) \mapsto f(s, \varphi_n(s, x))$ is also of class $C^1$, and thus one can use (1.71) together with Leibniz's rule to obtain the differentiability of $\varphi_{n+1}$ and identity (1.73). This implies that $\varphi_{n+1}$ is of class $C^1$.

Now we show that

$$\frac{\partial \varphi_n}{\partial x}(t, x) = \Phi_n(t, x) \tag{1.74}$$

for each $n \in \mathbb{N}$. Clearly,

$$\frac{\partial \varphi_1}{\partial x}(t, x) = \mathrm{Id} = \Phi_1(t, x).$$

In order to proceed by induction, we assume that (1.74) holds for a given $n$. It then follows from (1.73) and (1.72) that

$$\frac{\partial \varphi_{n+1}}{\partial x}(t, x) = \mathrm{Id} + \int_{t_0}^{t} \frac{\partial f}{\partial x}(s, \varphi_n(s, x)) \Phi_n(s, x) \, ds = \Phi_{n+1}(t, x).$$

This yields the desired identity.

Finally, given $t \in (t_0 - \alpha, t_0 + \alpha)$, we consider the sequence of functions $f_n(x) = \varphi_n(t, x)$. By (1.74), we have $d_x f_n = \Phi_n(t, x)$. On the other hand, as we already observed, by the Fiber contraction theorem (Theorem 1.38) the sequences $f_n$ and $df_n$ converge uniformly respectively to $\varphi_0(t, \cdot)$ and $\Phi_0(t, \cdot)$. Hence, it follows from Proposition 1.41 that the derivative $(\partial \varphi_0 / \partial x)(t, x)$ exists and satisfies (1.70). This completes the proof of the theorem. $\qquad\square$

**1.3.3. Maximal interval of existence.** In this section we show that each solution of the initial value problem (1.13) given by the Picard–Lindelöf theorem (Theorem 1.18) can be extended to a maximal interval in a unique manner.

**Theorem 1.43.** *If the function $f \colon D \to \mathbb{R}^n$ is continuous and locally Lipschitz in $x$ in an open set $D \subset \mathbb{R} \times \mathbb{R}^n$, then for each $(t_0, x_0) \in D$ there exists a unique solution $\varphi \colon (a, b) \to \mathbb{R}^n$ of the initial value problem (1.13) such that for any solution $x \colon I_x \to \mathbb{R}^n$ of the same problem we have $I_x \subset (a, b)$ and $x(t) = \varphi(t)$ for $t \in I_x$.*

**Proof.** We note that $J = \bigcup_x I_x$ is an open interval, because the union of any family of open intervals containing $t_0$ is still an open interval (containing $t_0$). Now we define a function $\varphi \colon J \to \mathbb{R}^n$ as follows. For each $t \in I_x$, let $\varphi(t) = x(t)$. We show that the function $\varphi$ is well defined, that is, $\varphi(t)$ does not depend on the function $x$. To that effect, let $x \colon I_x \to \mathbb{R}^n$ and $y \colon I_y \to \mathbb{R}^n$ be solutions of the initial value problem (1.13). Also, let $I$ be the largest open interval containing $t_0$ where $x = y$. We want to show that $I = I_x \cap I_y$. Otherwise, there would exist an endpoint $s$ of $I$ different from the endpoints of $I_x \cap I_y$. By the continuity of $x$ and $y$ in the interval $I_x \cap I_y$, we would have

$$p := \lim_{t \to s} x(t) = \lim_{t \to s} y(t).$$

On the other hand, by the Picard–Lindelöf theorem (Theorem 1.18) with $(t_0, x_0)$ replaced by $(s, p)$, there would exist an open interval $(s - \alpha, s + \alpha) \subset I_x \cap I_y$ where $x = y$. Since $(s - \alpha, s + \alpha) \setminus I \neq \varnothing$, this contradicts the fact that $I$ is the largest open interval containing $t_0$ where $x = y$. Therefore, $I = I_x \cap I_y$ and $x = y$ in $I_x \cap I_y$. Clearly, the function $\varphi \colon J \to \mathbb{R}^n$ is a solution of the initial value problem (1.13). This completes the proof of the theorem. □

In view of Theorem 1.43, one can introduce the following notion of maximal interval (of existence) of a solution.

**Definition 1.44.** Under the assumptions of Theorem 1.43, the *maximal interval* of a solution $x \colon I \to \mathbb{R}^n$ of the equation $x' = f(t, x)$ is the largest open interval $(a, b)$ where there exists a solution coinciding with $x$ in $I$.

We note that the maximal interval of a solution is not always $\mathbb{R}$.

**Example 1.45.** Consider the equation $x' = x^2$. One can easily verify (for example writing $x'/x^2 = 1$ and integrating on both sides) that the solutions are $x(t) = 0$, with maximal interval $\mathbb{R}$, and

$$x(t) = \frac{1}{c - t},$$

with maximal interval $(-\infty, c)$ or $(c, +\infty)$, for each $c \in \mathbb{R}$. More precisely, for $x_0 = 0$ the solution of the initial value problem

$$\begin{cases} x' = x^2, \\ x(t_0) = x_0 \end{cases}$$

is $x(t) = 0$, with maximal interval $\mathbb{R}$, for $x_0 > 0$ the solution is

$$x(t) = \frac{1}{t_0 - t + 1/x_0}, \tag{1.75}$$

with maximal interval $(-\infty, t_0 + 1/x_0)$ and, finally, for $x_0 < 0$ the solution is again given by (1.75), but now with maximal interval $(t_0 + 1/x_0, +\infty)$.

The following result describes what happens to the solutions of an ordinary differential equation when at least one endpoint of the maximal interval is a finite number.

**Theorem 1.46.** *Let $f \colon D \to \mathbb{R}^n$ be continuous and locally Lipschitz in $x$ in an open set $D \subset \mathbb{R} \times \mathbb{R}^n$. If a solution $x(t)$ of the equation $x' = f(t, x)$ has maximal interval $(a, b)$, then for each compact set $K \subset D$ there exists $\varepsilon > 0$ such that $(t, x(t)) \in D \setminus K$ for every*

$$t \in (a, a + \varepsilon) \cup (b - \varepsilon, b) \tag{1.76}$$

*(when $a = -\infty$ the first interval is empty and when $b = +\infty$ the second interval is empty).*

**Proof.** We consider only the endpoint $b$ (the argument for $a$ is entirely analogous). We proceed by contradiction. Let us assume that for some compact set $K \subset D$ there exists a sequence $(t_p)_p$ with $t_p \nearrow b$ when $p \to \infty$ such that

$$(t_p, x(t_p)) \in K \quad \text{for every} \quad p \in \mathbb{N}.$$

Since $K$ is compact, there exists a subsequence $(t_{k_p})_p$ such that $(t_{k_p}, x(t_{k_p}))$ converges to a point in $K$ when $p \to \infty$ (we recall that a set $K \subset \mathbb{R}^m$ is compact if and only if any sequence $(y_p)_p \subset K$ has a convergent subsequence with limit in $K$). Let

$$(b, x_0) = \lim_{p \to \infty} (t_{k_p}, x(t_{k_p})).$$

Now we consider the initial condition $x(b) = x_0$ and a compact set

$$K_{\alpha\beta} := [b - \alpha, b + \alpha] \times \overline{B(x_0, \beta)} \subset D$$

as in the proof of the Picard–Lindelöf theorem (Theorem 1.18), for some constants $\alpha, \beta > 0$ such that $2M\alpha \le \beta$, where

$$M = \sup \{\|f(t, x)\| : (t, x) \in K_{\alpha\beta}\}.$$

Moreover, for each $p \in \mathbb{N}$, we consider the compact set

$$L_p := [t_{k_p} - \alpha/2, t_{k_p} + \alpha/2] \times \overline{B(x(t_{k_p}), \beta/2)} \subset D.$$

For any sufficiently large $p$, we have $L_p \subset K_{\alpha\beta}$, and hence,

$$2 \sup \{\|f(t, x)\| : x \in L_p\} \alpha/2 \le 2M\alpha/2 \le \beta/2.$$

Thus, proceeding as in the proof of the Picard–Lindelöf theorem, we find that there exists a unique solution

$$y \colon \left(t_{k_p} - \alpha/2, t_{k_p} + \alpha/2\right) \to \mathbb{R}^n$$

of the equation $x' = f(t, x)$ with initial condition $y(t_{k_p}) = x(t_{k_p})$. Since $t_{k_p} + \alpha/2 > b$ for any sufficiently large $p$, this means that we obtain an extension of the solution $x$ to the interval $(a, t_{k_p} + \alpha/2)$. But this contradicts the fact that $b$ is the right endpoint of the maximal interval. Therefore, for each compact set $K \subset D$ there exists $\varepsilon > 0$ such that $(t, x(t)) \in D \setminus K$ for every $t \in (b - \varepsilon, b)$. $\square$

The following is an application of Theorem 1.46.

**Example 1.47.** Consider the equation in polar coordinates

$$\begin{cases} r' = -r, \\ \theta' = 2. \end{cases} \tag{1.77}$$

Proceeding as in (1.9) and (1.10), we obtain

$$x' = -r \cos \theta - 2r \sin \theta = -x - 2y$$

and

$$y' = -r \sin \theta + 2r \cos \theta = -y + 2x.$$

We observe that the assumptions of Theorem 1.46 are satisfied for the open set $D = \mathbb{R} \times \mathbb{R}^2$. Hence, for each compact set $K = [c,d] \times B \subset D$, where $B$ is a closed ball centered at the origin, there exists $\varepsilon > 0$ such that $(t, x(t), y(t)) \in D \setminus K$, that is,

$$t \notin [c,d] \quad \text{or} \quad (x(t), y(t)) \notin B, \tag{1.78}$$

for $t$ as in (1.76). On the other hand, by the first equation in (1.77), the distance of any solution to the origin decreases as time increases. This implies that each solution $(x(t), y(t))$ with initial condition $(x(t_0), y(t_0)) \in B$ remains in $B$ for every $t > t_0$. Hence, it follows from (1.78) that $t \notin [c,d]$ for any sufficiently large $t$ in the maximal interval of the solution. But since $d$ is arbitrary, we conclude that for any solution with initial condition in $B$ the right endpoint of its maximal interval is $+\infty$.

## 1.4. Existence of solutions for continuous fields

In this section we show that for a continuous function $f$ the initial value problem (1.13) always has solutions. However, these need not be unique (as illustrated by Example 1.23).

We first recall a result from functional analysis.

**Proposition 1.48** (Arzelà–Ascoli). *Let $\varphi_k \colon (a,b) \to \mathbb{R}^n$ be continuous functions, for $k \in \mathbb{N}$, and assume that:*

a) *there exists $c > 0$ such that*

$$\sup\{\|\varphi_k(t)\| : t \in (a,b)\} < c \quad \text{for every} \quad k \in \mathbb{N};$$

b) *given $\varepsilon > 0$, there exists $\delta > 0$ such that $\|\varphi_k(t) - \varphi_k(s)\| < \varepsilon$ for every $k \in \mathbb{N}$ and $t,s \in (a,b)$ with $|t - s| < \delta$.*

*Then there exists a subsequence of $(\varphi_k)_k$ converging uniformly to a continuous function in $(a,b)$.*

**Proof.** Consider a sequence $(t_m)_m \subset [a,b]$ that is dense in this interval. Since $(\varphi_k(t_1))_k$ is bounded, there exists a convergent subsequence $(\varphi_{p_k^1}(t_1))_k$. Similarly, since $(\varphi_{p_k^1}(t_2))_k$ is bounded, there exists a convergent subsequence $(\varphi_{p_k^2}(t_2))_k$. Proceeding always in this manner, after $m$ steps we obtain a convergent subsequence $(\varphi_{p_k^m}(t_m))_k$ of the bounded sequence $(\varphi_{p_k^{m-1}}(t_m))_k$. Now we define functions $\psi_k \colon (a,b) \to \mathbb{R}^n$ for $k \in \mathbb{N}$ by

$$\psi_k(t) = \varphi_{p_k^k}(t).$$

We note that the sequence $(\psi_k(t_m))_k$ is convergent for each $m \in \mathbb{N}$, because with the exception of the first $m$ terms it is a subsequence of the convergent sequence $(\varphi_{p_k^m}(t_m))_k$.

On the other hand, by hypothesis, for each $\varepsilon > 0$ there exists $\delta > 0$ such that

$$\|\psi_k(t) - \psi_k(s)\| = \|\varphi_{p_k^k}(t) - \varphi_{p_k^k}(s)\| < \varepsilon \tag{1.79}$$

for every $k \in \mathbb{N}$ and $t, s \in (a, b)$ with $|t - s| < \delta$. Given $t, t_m \in (a, b)$ with $|t - t_m| < \delta$ and $p, q \in \mathbb{N}$, it follows from (1.79) that

$$\begin{aligned}
\|\psi_p(t) - \psi_q(t)\| &\leq \|\psi_p(t) - \psi_p(t_m)\| \\
&\quad + \|\psi_p(t_m) - \psi_q(t_m)\| + \|\psi_q(t_m) - \psi_q(t)\| \\
&< 2\varepsilon + \|\psi_p(t_m) - \psi_q(t_m)\|.
\end{aligned} \tag{1.80}$$

Since the sequence $(\psi_k(t_m))_k$ is convergent, there exists $p_m \in \mathbb{N}$ such that

$$\|\psi_p(t_m) - \psi_q(t_m)\| < \varepsilon \tag{1.81}$$

for every $p, q \geq p_m$. Moreover, since the interval $(a, b)$ can be covered by a finite number $N$ of intervals of length $2\delta$, and without loss of generality centered at the points $t_1, t_2, \ldots, t_N$, it follows from (1.80) and (1.81) that

$$\|\psi_p(t) - \psi_q(t)\| < 3\varepsilon$$

for every $t \in (a, b)$ and $p, q \geq \max\{p_1, \ldots, p_N\}$. This shows that $(\psi_p)_p$ is a Cauchy sequence in the complete metric space $C(a, b)$ with the distance $d$ in (1.31) (see Proposition 1.30). Thus, the sequence converges uniformly to a continuous function in $(a, b)$. This completes the proof of the proposition. $\square$

Now we use Proposition 1.48 to establish the existence of solutions for any ordinary differential equation with a continuous right-hand side.

**Theorem 1.49** (Peano). *If $f: D \to \mathbb{R}^n$ is a continuous function in an open set $D \subset \mathbb{R} \times \mathbb{R}^n$, then for each $(t_0, x_0) \in D$ there exists at least one solution of the initial value problem* (1.13) *in some open interval containing $t_0$.*

**Proof.** We consider again the equivalent problem in (1.50), and we take $a < t_0 < b$ and $\beta > 0$ such that the conditions (1.51) and $(b - a)M \leq \beta$ hold, where $M$ is the constant in (1.52). Given $\alpha > 0$, we define recursively a continuous function $x_\alpha: (a, b) \to \mathbb{R}^n$ by

$$x_\alpha(t) = \begin{cases} x_0 - \int_t^{t_0 - \alpha} f(s, x_\alpha(s + \alpha)) \, ds, & t \in (a, t_0 - \alpha), \\ x_0, & t \in [t_0 - \alpha, t_0 + \alpha], \\ x_0 + \int_{t_0 + \alpha}^t f(s, x_\alpha(s - \alpha)) \, ds, & t \in [t_0 + \alpha, b). \end{cases} \tag{1.82}$$

For example,

$$x_\alpha(t) = x_0 + \int_{t_0+\alpha}^t f\big(s, x_\alpha(s-\alpha)\big)\, ds = x_0 + \int_{t_0+\alpha}^t f(s, x_0)\, ds$$

for $t \in [t_0 + \alpha, t_0 + 2\alpha]$, and

$$x_\alpha(t) = x_0 - \int_t^{t_0-\alpha} f\big(s, x_\alpha(s+\alpha)\big)\, ds = x_0 + \int_{t_0-\alpha}^t f(s, x_0)\, ds$$

for $t \in [t_0 - 2\alpha, t_0 - \alpha]$. We have

$$\|x_\alpha(t) - x_0\| \le \left| \int_{t_0 \pm \alpha}^t \big\| f\big(s, x_\alpha(s \mp \alpha)\big)\big\|\, ds \right| \le (b-a)M \le \beta,$$

and thus,

$$\|x_\alpha(t)\| \le \|x_0\| + \beta.$$

Moreover,

$$\|x_\alpha(t) - x_\alpha(s)\| \le \left| \int_s^t \big\| f\big(u, x_\alpha(u \mp \alpha)\big)\big\|\, du \right| \le M|t - s|. \qquad (1.83)$$

This shows that the two conditions in Proposition 1.48 are satisfied for any sequence of functions $(x_{\alpha_m})_m$, where $(\alpha_m)_m \subset \mathbb{R}^+$ is an arbitrary sequence such that $\alpha_m \searrow 0$ when $m \to \infty$. Therefore, there exists a subsequence $(\beta_m)_m$ of $(\alpha_m)_m$ such that $(x_{\beta_m})_m$ converges uniformly to a continuous function $x \colon (a, b) \to \mathbb{R}^n$.

It remains to show that $x$ is a solution of the initial value problem (1.13). By (1.83), we have

$$\big\| x_{\beta_m}(s - \beta_m) - x(s) \big\| \le \big\| x_{\beta_m}(s - \beta_m) - x_{\beta_m}(s) \big\| + \big\| x_{\beta_m}(s) - x(s) \big\|$$
$$\le M|\beta_m| + \|x_{\beta_m}(s) - x(s)\|,$$

and thus, the sequence of functions $s \mapsto x_{\beta_m}(s - \beta_m)$ converges uniformly to $x$. One can show in a similar manner that the sequence of functions $s \mapsto x_{\beta_m}(s + \beta_m)$ also converges uniformly to $x$. On the other hand, it follows from (1.82) that

$$x_{\beta_m}(t) = x_0 + \int_{t_0+\beta_m}^t f\big(s, x_{\beta_m}(s - \beta_m)\big)\, ds$$

for $t \in [t_0, b)$, and letting $m \to \infty$, we obtain

$$x(t) = x_0 + \int_{t_0}^t f(s, x(s))\, ds \qquad (1.84)$$

for $t \in [t_0, b)$. Analogously, it follows from (1.82) that

$$x_{\beta_m}(t) = x_0 - \int_t^{t_0-\beta_m} f\big(s, x_{\beta_m}(s + \beta_m)\big)\, ds$$

for $t \in (a, t_0]$, and letting $m \to \infty$, we obtain identity (1.84) for $t \in (a, t_0]$. Finally, it follows from Proposition 1.9 that $x$ is a solution of the initial value problem (1.13). □

## 1.5. Phase portraits

Although it is often impossible (or very difficult) to determine explicitly the solutions of an ordinary differential equation, it is still important to obtain information about these solutions, at least of qualitative nature. To a considerable extent, this can be done describing the phase portrait of the differential equation. We note that in this section we consider only autonomous equations.

**1.5.1. Orbits.** Let $f \colon D \to \mathbb{R}^n$ be a continuous function in an open set $D \subset \mathbb{R}^n$, and consider the autonomous equation

$$x' = f(x). \tag{1.85}$$

The set $D$ is called the *phase space* of the equation.

**Definition 1.50.** If $x = x(t)$ is a solution of equation (1.85) with maximal interval $I$, then the set $\{x(t) : t \in I\} \subset D$ is called an *orbit* of the equation.

We first consider a very simple class of orbits.

**Definition 1.51.** A point $x_0 \in D$ with $f(x_0) = 0$ is called a *critical point* of equation (1.85).

**Proposition 1.52.** *If $x_0$ is a critical point of equation* (1.85)*, then $\{x_0\}$ is an orbit of this equation.*

**Proof.** It is sufficient to observe that the constant function $x \colon \mathbb{R} \to \mathbb{R}^n$ defined by $x(t) = x_0$ is a solution of equation (1.85). □

**Example 1.53.** By Example 1.45, the equation $x' = x^2$ has the solutions $x(t) = 0$, with maximal interval $\mathbb{R}$, and $x(t) = 1/(c - t)$, with maximal interval $(-\infty, c)$ or $(c, +\infty)$, for each $c \in \mathbb{R}$. Thus, we obtain the orbits $\{0\}$ (coming from the critical point 0),

$$\{1/(c - t) : t \in (-\infty, c)\} = \mathbb{R}^+$$

and

$$\{1/(c - t) : t \in (c, +\infty)\} = \mathbb{R}^-.$$

Now we show that in a sufficiently small neighborhood of a noncritical point one can always introduce coordinates with respect to which the orbits are parallel line segments traversed with constant speed.

**Theorem 1.54** (Flow box theorem). *Let* $f \colon D \to \mathbb{R}^n$ *be a function of class* $C^1$ *in an open set* $D \subset \mathbb{R}^n$. *Given a point* $p \in D$ *with* $f(p) \neq 0$, *there exists a coordinate change* $y = g(x)$ *in a neighborhood of* $p$ *transforming the equation* $x' = f(x)$ *into the equation* $y' = v$ *for some* $v \in \mathbb{R}^n \setminus \{0\}$.

**Proof.** Let $\varphi_t(x_0)$ be the solution of the equation $x' = f(x)$ with initial condition $x(0) = x_0$, in its maximal interval. Since $f(p) \neq 0$, some component of $f(p)$ is different from zero. Without loss of generality, we assume that the first component is nonzero, and we define new coordinates $y = (y_1, \ldots, y_n)$ by

$$x = F(y) = \varphi_{y_1}(p + \bar{y}),$$

where $p = (p_1, \ldots, p_n)$ and $\bar{y} = (0, y_2, y_3, \ldots, y_n)$. By Theorem 1.42, the function $F$ is of class $C^1$ in some neighborhood of zero. Moreover, $F(0) = p$. Now we compute the Jacobian matrix $d_0 F$. The first column of $d_0 F$ is given by

$$\frac{\partial}{\partial y_1} \varphi_{y_1}(p + \bar{y})\big|_{y=0} = f\big(\varphi_{y_1}(p + \bar{y})\big)\big|_{y=0} = f(F(0)) = f(p).$$

Since $\varphi_0 = \mathrm{Id}$, the remaining columns are

$$\frac{\partial}{\partial y_i} \varphi_{y_1}(p + \bar{y}) = \frac{\partial}{\partial y_i} \varphi_{y_1}(p_1, p_2 + y_2, p_3 + y_3, \ldots, p_n + y_n)\big|_{y=0} = e_i$$

for $i = 2, \ldots, n$, where $(e_1, \ldots, e_n)$ is the canonical basis of $\mathbb{R}^n$. Hence,

$$d_0 F = (f(p) e_2 \cdots e_n)$$

is a lower triangular matrix with nonzero entries in the diagonal (because by hypothesis the first component of $f(p)$ is nonzero). Thus, $d_0 F$ is nonsingular and $F$ is indeed a coordinate change in a neighborhood of zero. Now let

$$(x_1(t), \ldots, x_n(t)) = \varphi_t(p + q)$$

be a solution of the equation $x' = f(x)$, where $q = (0, q_2, q_3, \ldots, q_n)$, and write it in the new coordinates in the form $y(t) = (y_1(t), \ldots, y_n(t))$. Since

$$\varphi_t(p + q) = F(y(t)) = \varphi_{y_1(t)}(p_1, p_2 + y_2(t), \ldots, p_n + y_n(t)),$$

we conclude that

$$y(t) = (t, q_2, q_3, \ldots, q_n).$$

Taking derivatives, we finally obtain $y' = (1, 0, \ldots, 0)$.  $\square$

Now we consider other types of orbits (see Figure 1.2).

**Definition 1.55.** An orbit $\{x(t) : t \in (a, b)\}$ that is not a critical point is said to be:

   a) *periodic* if there exists $T > 0$ such that $x(t + T) = x(t)$ whenever $t, t + T \in (a, b)$;

b) *homoclinic* if there exists a critical point $p$ such that $x(t) \to p$ when $t \to a^+$ and when $t \to b^-$;

c) *heteroclinic* if there exist critical points $p$ and $q$ with $p \neq q$ such that $x(t) \to p$ when $t \to a^+$ and $x(t) \to q$ when $t \to b^-$.

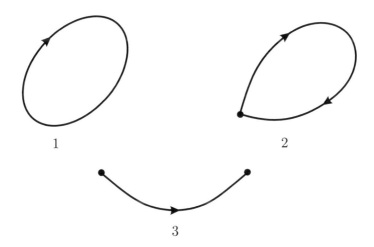

**Figure 1.2.** 1) periodic orbit; 2) homoclinic orbit; 3) heteroclinic orbit.

We show that all orbits in Definition 1.55 and more generally all orbits contained in a compact set have maximal interval $\mathbb{R}$.

**Definition 1.56.** A solution $x = x(t)$ is said to be *global* if its maximal interval is $\mathbb{R}$.

**Proposition 1.57.** *Any solution of equation* (1.85) *whose orbit is contained in a compact subset of $D$ is global.*

**Proof.** Let $F \colon \mathbb{R} \times D \to \mathbb{R}^n$ be the function $F(t,x) = f(x)$. Equation (1.85) can be written in the form

$$x' = F(t,x).$$

Given an orbit $\{x(t) : t \in (a,b)\}$ contained in some compact set $K \subset D$, let us consider the compact set $[-m,m] \times K \subset \mathbb{R} \times D$. It follows from Theorem 1.46 that there exists $\varepsilon > 0$ such that

$$(t, x(t)) \notin [-m,m] \times K$$

for every $t \in (a, a + \varepsilon) \cup (b - \varepsilon, b)$. Hence, there exists $\varepsilon > 0$ such that $t \notin [-m,m]$ for the same values of $t$ (since the orbit is contained in $K$). Letting $m \to \infty$, we conclude that $a = -\infty$ and $b = +\infty$. $\square$

The following is a consequence of Proposition 1.57.

**Proposition 1.58.** *Critical points, periodic orbits, homoclinic orbits and heteroclinic orbits are always obtained from global solutions.*

**1.5.2. Phase portraits.** The *phase portrait* of an autonomous ordinary differential equation is obtained representing the orbits in the set $D$, also indicating the direction of motion. It is common not to indicate the directions of the axes, since these could be confused with the direction of motion.

**Example 1.59** (continuation of Example 1.53). We already know that the orbits of the equation $x' = x^2$ are $\{0\}$, $\mathbb{R}^+$ and $\mathbb{R}^-$. These give rise to the phase portrait in Figure 1.3, where we also indicated the direction of motion. We note that in order to determine the phase portrait it is not necessary to solve the equation explicitly.

**Figure 1.3.** Phase portrait of the equation $x' = x^2$.

**Example 1.60.** Consider the equation $x' = |x|$. The only critical point is the origin. Moreover, for $x \neq 0$ we have $x' = |x| > 0$, and hence, all solutions avoiding the origin are increasing. Again, this yields the phase portrait in Figure 1.3, that is, the phase portraits of the equations $x' = x^2$ and $x' = |x|$ are the same. However, the speed along the orbits is not the same in the two equations.

Examples 1.59 and 1.60 illustrate that different equations can have the same phase portrait or, in other words, the same qualitative behavior. We emphasize that phase portraits give no quantitative information.

**Example 1.61.** Consider the equation

$$(x, y)' = (y, -x) \tag{1.86}$$

(see Example 1.3). The only critical point is $(0, 0)$. Moreover, by (1.6), if $(x, y)$ is a solution, then $(x^2 + y^2)' = 0$, and thus, other than the origin all orbits are circles. In order to determine the direction of motion it is sufficient to consider any point of each orbit. For example, at $(x, y) = (x, 0)$ we have $(x, y)' = (0, -x)$, and thus, the phase portrait is the one shown in Figure 1.4. We note that it was not necessary to use the explicit form of the solutions in (1.7).

It is sometimes useful to write the equation in other coordinates, as illustrated by the following example.

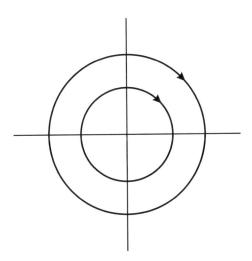

**Figure 1.4.** Phase portrait of the equation $(x, y)' = (y, -x)$.

**Example 1.62** (continuation of Example 1.61). Let us write equation (1.86) in polar coordinates. Since $x = r \cos \theta$ and $y = r \sin \theta$, we have

$$r' = \left(\sqrt{x^2 + y^2}\right)' = \frac{(x^2 + y^2)'}{2\sqrt{x^2 + y^2}} = \frac{xx' + yy'}{r}$$

and

$$\theta' = \left(\arctan \frac{y}{x}\right)' = \frac{(y/x)'}{1 + (y/x)^2} = \frac{y'x - x'y}{r^2}.$$

Thus, equation (1.86) takes the form

$$\begin{cases} r' = 0, \\ \theta' = -1 \end{cases}$$

in polar coordinates. Since $r' = 0$, all solutions remain at a constant distance from the origin, traversing circles (centered at the origin) with angular velocity $\theta' = -1$. Thus, the phase portrait is the one shown in Figure 1.4.

**Example 1.63.** Consider the equation

$$(x, y)' = (y, x). \tag{1.87}$$

The only critical point is the origin. Moreover, if $(x, y)$ is a solution, then

$$(x^2 - y^2)' = 2xx' - 2yy' = 2xy - 2yx = 0.$$

This shows that each orbit is contained in one of the hyperbolas defined by $x^2 - y^2 = c$, for some constant $c \neq 0$, or in one of the straight lines bisecting the quadrants, namely $x = y$ and $x = -y$. In order to determine the direction of motion in each orbit, it is sufficient to observe, for example, that $x' > 0$ for $y > 0$, and that $x' < 0$ for $y < 0$. The phase portrait is the one shown in Figure 1.5.

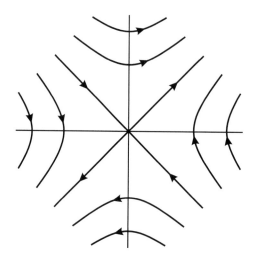

**Figure 1.5.** Phase portrait of the equation $(x, y)' = (y, x)$.

**Example 1.64.** Consider the equation

$$\begin{cases} x' = y(x^2 + 1), \\ y' = x(x^2 + 1). \end{cases} \tag{1.88}$$

As in Example 1.63, the origin is the only critical point, and along the solutions we have

$$(x^2 - y^2)' = 2xx' - 2yy'$$
$$= 2xy(x^2 + 1) - 2yx(x^2 + 1) = 0.$$

Since the signs of $y(x^2+1)$ and $x(x^2+1)$ coincide respectively with the signs of $y$ and $x$, the phase portrait of equation (1.88) is the same as the phase portrait of equation (1.87) (see Figure 1.5).

**Example 1.65.** Consider the equation

$$\begin{cases} x' = y(x^2 - 1), \\ y' = x(x^2 - 1). \end{cases} \tag{1.89}$$

The critical points are obtained solving the system

$$y(x^2 - 1) = x(x^2 - 1) = 0,$$

whose solutions are $(0,0)$, $(1,y)$ and $(-1,y)$, with $y \in \mathbb{R}$. Moreover, along the solutions we have

$$(x^2 - y^2)' = 2xx' - 2yy'$$
$$= 2xy(x^2 - 1) - 2yx(x^2 - 1) = 0,$$

and thus, each orbit is again contained in one of the hyperbolas defined by $x^2 - y^2 = c$, for some constant $c \neq 0$, or in one of the straight lines $x = y$ and $x = -y$. The phase portrait of equation (1.89) is the one shown in Figure 1.6, where the direction of motion can be obtained, for example, from the sign of $y' = x(x^2 - 1)$; namely,

$$y' > 0 \quad \text{for} \quad x \in (-1, 0) \cup (1, +\infty)$$

and

$$y' < 0 \quad \text{for} \quad x \in (-\infty, -1) \cup (0, 1).$$

In particular, it follows from the phase portrait that any orbit intersecting the vertical strip

$$L = \big\{(x, y) \in \mathbb{R}^2 : -1 \leq x \leq 1\big\}$$

is contained in $L$ and is bounded, while any other orbit is unbounded.

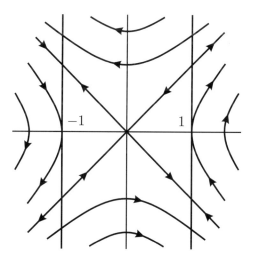

**Figure 1.6.** Phase portrait of equation (1.89).

The following example uses complex variables.

**Example 1.66.** Consider the equation

$$\begin{cases} x' = x^2 - y^2, \\ y' = 2xy. \end{cases} \tag{1.90}$$

Letting $z = x + iy$, one can easily verify that equation (1.90) is equivalent to $z' = z^2$, where $z' = x' + iy'$. Dividing by $z^2$ yields the equation

$$\frac{z'}{z^2} = \left(-\frac{1}{z}\right)' = 1,$$

whose solutions are

$$z(t) = -\frac{1}{t + c}, \tag{1.91}$$

with $c \in \mathbb{C}$. In order to obtain $x(t)$ and $y(t)$, it remains to consider the real and imaginary parts of $z(t)$ in (1.91). Namely, if $c = a + ib$ with $a, b \in \mathbb{R}$, then

$$(x(t), y(t)) = \frac{(-a - t, b)}{(a + t)^2 + b^2}.$$

One can use these formulas to verify that the phase portrait is the one shown in Figure 1.7.

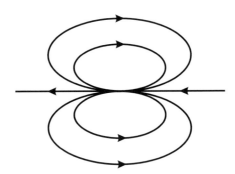

**Figure 1.7.** Phase portrait of equation (1.90).

**1.5.3. Conservative equations.** In this section we consider the particular case of the differential equations conserving a given function. We first introduce the notion of conservative equation.

**Definition 1.67.** A function $E \colon D \to \mathbb{R}$ of class $C^1$ that is constant in no open set and that is constant along the solutions of the equation $x' = f(x)$ is said to be an *integral* (or a *first integral*) of this equation. When there exists an integral the equation is said to be *conservative*.

We note that a function $E$ of class $C^1$ is constant along the solutions of the equation $x' = f(x)$ if and only if

$$\frac{d}{dt} E(x(t)) = 0$$

for any solution $x = x(t)$. More generally, one can consider integrals in subsets of $D$ and integrals that are not of class $C^1$.

**Example 1.68.** It follows from identity (1.6) that the function $E \colon \mathbb{R}^2 \to \mathbb{R}$ defined by $E(x,y) = x^2 + y^2$ is an integral of equation (1.4), and thus, the equation is conservative.

**Example 1.69.** Let $E \colon D \to \mathbb{R}$ be a function of class $C^1$ that is constant in no open set. Then the equation in $D \times D \subset \mathbb{R}^{2n}$ given by

$$\begin{cases} x' = \partial E / \partial y, \\ y' = -\partial E / \partial x \end{cases} \tag{1.92}$$

is conservative and the function $E$ is an integral. Indeed, if $(x,y)$ is a solution of equation (1.92), then

$$\begin{aligned} \frac{d}{dt} E(x,y) &= \frac{\partial E}{\partial x} x' + \frac{\partial E}{\partial y} y' \\ &= -y'x' + x'y' = 0, \end{aligned} \tag{1.93}$$

which shows that $E$ is an integral. Incidentally, equation (1.4) is obtained from (1.92) taking

$$E(x,y) = \frac{x^2 + y^2}{2}.$$

**Example 1.70.** Now we consider equation (1.92) with

$$E(x,y) = xy(x + y - 1), \tag{1.94}$$

which can be written in the form

$$\begin{cases} x' = x^2 + 2xy - x, \\ y' = -2xy - y^2 + y. \end{cases} \tag{1.95}$$

In view of (1.93), in order to obtain the phase portrait of equation (1.95) we have to determine the level sets of the function $E$ in (1.94). These are sketched in Figure 1.8, where we have also indicated the direction of motion.

One can easily verify that the critical points of equation (1.95) are $(0,0)$, $(1,0)$, $(0,1)$ and $(1/3, 1/3)$. Now we show that the orbits in the interior of the triangle determined by $(0,0)$, $(1,0)$ and $(0,1)$, in a neighborhood of the critical point $(1/3, 1/3)$, are in fact periodic orbits. Writing $X = x - 1/3$ and $Y = y - 1/3$, we obtain

$$\begin{aligned} E(x,y) &= \left( X + \frac{1}{3} \right) \left( Y + \frac{1}{3} \right) \left( X + Y - \frac{1}{3} \right) \\ &= -\frac{1}{27} + \frac{1}{3}(X^2 + XY + Y^2) + (X + Y)XY. \end{aligned}$$

Since the quadratic form

$$X^2 + XY + Y^2 = \begin{pmatrix} X \\ Y \end{pmatrix}^* \begin{pmatrix} 1 & 1/2 \\ 1/2 & 1 \end{pmatrix} \begin{pmatrix} X \\ Y \end{pmatrix} \tag{1.96}$$

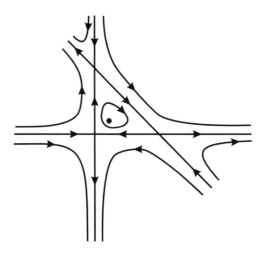

**Figure 1.8.** Phase portrait of equation (1.95).

is positive definite (the $2 \times 2$ matrix in (1.96) has eigenvalues $1/2$ and $3/2$), it follows from Morse's lemma (Proposition 9.16) that there is a coordinate change $(\bar{X}, \bar{Y}) = g(X, Y)$ in neighborhoods of $(0, 0)$ such that

$$E(x, y) = -\frac{1}{27} + \frac{1}{3}(\bar{X}^2 + \bar{X}\bar{Y} + \bar{Y}^2).$$

Since the level sets of $E$ are closed curves in the variables $(\bar{X}, \bar{Y})$, in a neighborhood of $(0, 0)$, the same happens in the variables $(x, y)$, now in a neighborhood of $(1/3, 1/3)$.

**Example 1.71.** Consider the equation

$$\begin{cases} x' = y, \\ y' = f(x), \end{cases} \tag{1.97}$$

where

$$f(x) = x(x - 1)(x - 3). \tag{1.98}$$

The critical points are $(0, 0)$, $(1, 0)$ and $(3, 0)$. Now we consider the function

$$E(x, y) = \frac{1}{2}y^2 - \int_0^x f(s)\, ds$$

$$= \frac{1}{2}y^2 - \frac{1}{4}x^4 + \frac{4}{3}x^3 - \frac{3}{2}x^2.$$

If $(x, y) = (x(t), y(t))$ is a solution of equation (1.97), then

$$\frac{d}{dt}E(x, y) = yy' - f(x)x' = yf(x) - f(x)y = 0, \tag{1.99}$$

which shows that the function $E$ is an integral. In order to obtain the phase portrait, we have to determine the level sets of the function $E$. One can show that these are the curves sketched in Figure 1.9, which determines the phase portrait. In particular, there exists a homoclinic orbit connecting the origin to itself.

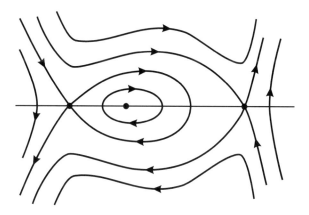

**Figure 1.9.** Phase portrait of equation (1.97) with $f$ given by (1.98).

In order to determine the level sets of the function $E$, we proceed as follows. We first write equation (1.97) in the form

$$x'' = y' = f(x).$$

This corresponds to apply a force $f(x)$ to a particle of mass 1 moving without friction. For example, $f(x)$ can be the gravitational force resulting from the particle being at a height $f(x)$. Actually, in order to obtain the phase portrait it is not important what $f(x)$ really is, and so there is no loss of generality in assuming that $f(x)$ is indeed the gravitational force. In this context, identity (1.99) corresponds to the conservation of the total energy $E(x, y)$, which is the sum of the kinetic energy $y^2/2$ with the potential energy

$$V(x) = -\int_0^x f(s)\, ds = -\frac{1}{4}x^4 + \frac{4}{3}x^3 - \frac{3}{2}x^2.$$

The function $V$ has the graph shown in Figure 1.10, which can be obtained noting that

$$V'(x) = -x(x-1)(x-3).$$

The phase portrait can now be obtained from the energy conservation. For example, it follows from Figure 1.10 that a ball dropped with velocity $y = 0$ at a point $x$ in the interval $(0, (8-\sqrt{10})/3)$ starts descending and after some time passes through the point $x = 1$, where it attains maximal speed. Then it starts moving upward, in the same direction, up to a point $x' \in (0, (8-\sqrt{10})/3)$ with $V(x') = V(x)$, where again it attains the velocity $y = 0$. This type of behavior repeats itself indefinitely, with the ball oscillating

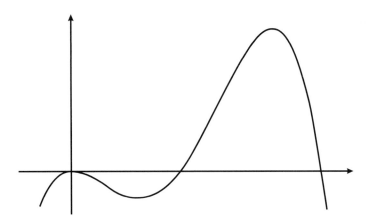

**Figure 1.10.** Graph of the function $V$ in Example 1.71. The points of intersection with the horizontal axis are $0$, $(8-\sqrt{10})/3$ and $(8+\sqrt{10})/3$.

back and forth between the points $x$ and $x'$. This corresponds to one of the periodic orbits in Figure 1.9, in the interior of the homoclinic orbit. On the other hand, a ball dropped with velocity $y > 0$ at a point $x \in (0,1)$ with $y^2/2 + V(x) = 0$ (by Figure 1.10 the function $V$ is negative in the interval $(0,1)$) travels along the same path up to a point $x'$ with $V(x') = V(x)$, although now without zero velocity, because of the conservation of energy. Instead, it continues traveling up to a point $x'' \in (1, +\infty)$ with $y^2 + V(x) = V(x'')$, where it finally attains zero velocity (the existence of this point follows from Figure 1.10 and the conservation of energy). Then the ball starts traveling backwards, in particular, passing through the initial point $x$, and approaching the origin indefinitely, without ever getting there, since

$$y^2 + V(x) = 0 = E(0,0).$$

Analogously, a ball dropped with velocity $y < 0$ at a point $x \in (0,1)$ with $y^2/2 + V(x) = 0$ exhibits the same type of behavior. Namely, it approaches the origin indefinitely, without ever getting there. This corresponds to the homoclinic orbit in Figure 1.10. The remaining orbits can be described in an analogous manner.

**Example 1.72.** Now we consider equation (1.97) with

$$f(x) = x(x-1)(x-2). \tag{1.100}$$

The critical points are $(0,0)$, $(1,0)$ and $(2,0)$. Moreover, the function

$$\begin{aligned} E(x,y) &= \frac{1}{2}y^2 - \int_0^x f(s)\,ds \\ &= \frac{1}{2}y^2 - \frac{1}{4}x^4 + x^3 - x^2 \end{aligned} \tag{1.101}$$

is an integral of equation (1.97). The phase portrait is the one shown in Figure 1.11. In particular, there exist two heteroclinic orbits connecting the critical points $(0,0)$ and $(2,0)$.

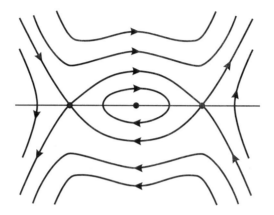

**Figure 1.11.** Phase portrait of equation (1.97) with $f$ given by (1.100).

In order to determine the level sets of the function $E$ in (1.101), we proceed in a similar manner to that in Example 1.71. In this case the potential energy is given by

$$V(x) = -\int_0^x f(s)\,ds = -\frac{1}{4}x^4 + x^3 - x^2,$$

and has the graph shown in Figure 1.12.

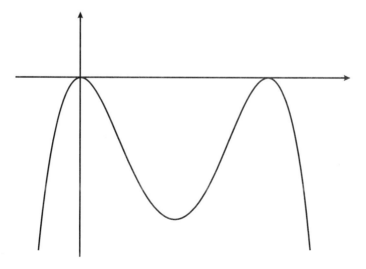

**Figure 1.12.** Graph of the function $V$ in Example 1.72. The points of intersection with the horizontal axis are 0 and 2.

Now we use the conservation of energy to describe the phase portrait. In particular, for a ball dropped with velocity $y = 0$ at a point $x \in (0, 2)$, we obtain an oscillatory movement between $x$ and $2 - x$. Indeed,

$$V(x) = -\frac{1}{4}x^4 + x^3 - x^2 = -\frac{1}{4}x^2(x-2)^2,$$

and hence $V(2 - x) = V(x)$. This corresponds to one of the periodic orbits in Figure 1.11. On the other hand, a ball dropped with velocity $y > 0$ at a point $x \in (0, 2)$ with energy $y^2/2 + V(x) = 0$ approaches indefinitely the point 2, without ever getting there, since $E(2, 0) = 0$. Also, considering the movement for negative time, we obtain the heteroclinic orbit in the upper half-plane in Figure 1.11. The heteroclinic orbit in the lower half-plane corresponds to the trajectory of a ball dropped with velocity $y < 0$ at a point $x \in (0, 2)$ with energy $y^2/2 + V(x) = 0$. The remaining orbits in Figure 1.11 can be described in an analogous manner.

## 1.6. Equations on manifolds

This section is a brief introduction to the study of ordinary differential equations on (smooth) manifolds. After recalling the relevant notions concerning manifolds and vector fields, we consider ordinary differential equations defined by vector fields on manifolds and we give several examples.

**1.6.1. Basic notions.** In this section we recall some basic notions concerning manifolds. We first introduce the notion of a differentiable structure.

**Definition 1.73.** A set $M$ is said to admit a *differentiable structure* of class $C^k$ and dimension $n$ if there exist injective maps $\varphi_i \colon U_i \to M$ defined in open sets $U_i \subset \mathbb{R}^n$ for $i \in I$ such that:

a) $\bigcup_{i \in I} \varphi_i(U_i) = M$;

b) for any $i, j \in I$ with $V = \varphi_i(U_i) \cap \varphi_j(U_j) \neq \varnothing$ the preimages $\varphi_i^{-1}(V)$ and $\varphi_j^{-1}(V)$ are open sets in $\mathbb{R}^n$, and the map $\varphi_j^{-1} \circ \varphi_i$ is of class $C^k$.

It is sometimes required that the family $(U_i, \varphi_i)_{i \in I}$ is maximal with respect to the conditions in Definition 1.73, although one can always complete it to a maximal one. Each map $\varphi_i \colon U_i \to M$ is called a *chart* or a *coordinate system*. Any differentiable structure induces a topology on $M$; namely, a set $A \subset M$ is said to be *open* if $\varphi_i^{-1}A \subset \mathbb{R}^n$ is an open set for every $i \in I$.

Now we introduce the notion of a manifold.

**Definition 1.74.** A set $M$ is said to be a *manifold* of class $C^k$ and dimension $n$ if:

a) $M$ admits a differentiable structure of class $C^k$ and dimension $n$;

b) when equipped with the corresponding induced topology, the set $M$ is a Hausdorff topological space with countable basis.

We recall that a topological space is said to be Hausdorff if distinct points have disjoint open neighborhoods, and that it is said to have a countable basis if any open set can be written as a union of elements of a given countable family of open sets.

**Example 1.75.** The real line $\mathbb{R}$ is a manifold of dimension 1 when equipped with the differentiable structure composed of the single chart $\varphi\colon \mathbb{R} \to \mathbb{R}$ given by $\varphi(x) = x$.

**Example 1.76.** The circle

$$S^1 = \left\{ (x, y) \in \mathbb{R}^2 : x^2 + y^2 = 1 \right\}$$

is a manifold of dimension 1. A differentiable structure is given by the maps

$$\varphi_i\colon (-1, 1) \to S^1, \quad i = 1, 2, 3, 4$$

defined by

$$
\begin{aligned}
\varphi_1(x) &= \left( x, \sqrt{1 - x^2} \right), \\
\varphi_2(x) &= \left( x, -\sqrt{1 - x^2} \right), \\
\varphi_3(x) &= \left( \sqrt{1 - x^2}, x \right), \\
\varphi_4(x) &= \left( -\sqrt{1 - x^2}, x \right).
\end{aligned}
\tag{1.102}
$$

One can show in a similar manner that the $n$-sphere

$$S^n = \left\{ x \in \mathbb{R}^{n+1} : \|x\| = 1 \right\},$$

where

$$\|(x_1, \ldots, x_{n+1})\| = \left( \sum_{i=1}^{n+1} x_i^2 \right)^{1/2},$$

is a manifold of dimension $n$.

**Example 1.77.** The $n$-torus $\mathbb{T}^n = S^1 \times \cdots \times S^1$ is a manifold of dimension $n$. A differentiable structure is given by the maps $\psi\colon (-1, 1)^n \to \mathbb{T}^n$ defined by

$$\psi(x_1, \ldots, x_n) = \left( \psi_1(x_1), \ldots, \psi_n(x_n) \right),$$

where each $\psi_i$ is any of the functions $\varphi_1$, $\varphi_2$, $\varphi_3$ and $\varphi_4$ in (1.102).

**Example 1.78.** Let $\varphi\colon U \to \mathbb{R}^m$ be a function of class $C^k$ in an open set $U \subset \mathbb{R}^n$. Then the graph

$$M = \left\{ (x, \varphi(x)) : x \in U \right\} \subset \mathbb{R}^n \times \mathbb{R}^m$$

is a manifold of class $C^k$ and dimension $n$. A differentiable structure is given by the map $\psi\colon U \to \mathbb{R}^n \times \mathbb{R}^m$ defined by $\psi(x) = (x, \varphi(x))$.

**Example 1.79.** Let $E\colon U \to \mathbb{R}^m$ be a function of class $C^1$ in an open set $U \subset \mathbb{R}^n$. A point $c \in \mathbb{R}^m$ is said to be a *critical value* of $E$ if $c = E(x)$ for some $x \in U$ such that $d_x F\colon \mathbb{R}^n \to \mathbb{R}^m$ is not onto. One can use the Inverse function theorem to show that if $c \in \mathbb{R}^m$ is not a critical value of $E$, then the level set

$$E^{-1}c = \{x \in U : E(x) = c\}$$

is a manifold of dimension $n - m$.

For example, if $E\colon U \to \mathbb{R}$ is an integral of a conservative equation $x' = f(x)$, then for each $c \in \mathbb{R}$ that is not a critical value of $E$, the level set $E^{-1}c$ is a manifold of dimension $n - 1$.

**1.6.2. Vector fields.** In order to introduce the notion of a tangent vector, we first recall the notion of a differentiable map between manifolds.

**Definition 1.80.** A map $f\colon M \to N$ between manifolds $M$ and $N$ is said to be *differentiable* at a point $x \in M$ if there exist charts $\varphi\colon U \to M$ and $\psi\colon V \to N$ such that:

    a) $x \in \varphi(U)$ and $f(\varphi(U)) \subset \psi(V)$;

    b) $\psi^{-1} \circ f \circ \varphi$ is differentiable at $\varphi^{-1}(x)$.

We also say that $f$ is of class $C^k$ in an open set $W \subset M$ if all maps $\psi^{-1} \circ f \circ \varphi$ are of class $C^k$ in $\varphi^{-1}(W)$.

Now we introduce the notion of a tangent vector to a curve. Let $M$ be a manifold of dimension $n$. A differentiable function $\alpha\colon (-\varepsilon, \varepsilon) \to M$ is called a *curve*. Also, let $D$ be the set of all functions $f\colon M \to \mathbb{R}$ that are differentiable at a given point $x \in M$.

**Definition 1.81.** The *tangent vector* to a curve $\alpha\colon (-\varepsilon, \varepsilon) \to M$ with $\alpha(0) = x$ at $t = 0$ is the function $F\colon D \to \mathbb{R}$ defined by

$$F(f) = \frac{d(f \circ \alpha)}{dt}\bigg|_{t=0}.$$

We also say that $F$ is a *tangent vector* at $x$.

One can easily verify that the set $T_x M$ of all tangent vectors at $x$ is a vector space of dimension $n$. It is called the *tangent space* of $M$ at $x$. The *tangent bundle* of $M$ is the set

$$TM = \{(x, v) : x \in M, v \in T_x M\}.$$

One can show that $TM$ is a manifold of dimension $2n$. A differentiable structure can be obtained as follows. Let $\varphi\colon U \to M$ be a chart for $M$ and let $(x_1, \dots, x_n)$ be the coordinates in $U$. Consider the curves $\alpha_i\colon (-\varepsilon, \varepsilon) \to M$ for $i = 1, \dots, n$ defined by

$$\alpha_i(t) = \varphi(te_i),$$

where $(e_1, \ldots, e_n)$ is the standard basis of $\mathbb{R}^n$. The tangent vector to the curve $\alpha_i$ at $t = 0$ is denoted by $\partial/\partial x_i$. One can easily verify that $(\partial/\partial x_1, \ldots, \partial/\partial x_n)$ is a basis of the tangent space $T_{\varphi(0)}M$. Now we consider the map $\psi \colon U \times \mathbb{R}^n \to TM$ defined by

$$\psi(x_1, \ldots, x_n, y_1, \ldots, y_n) = \left( \varphi(x_1, \ldots, x_n), \sum_{i=1}^{n} y_i \frac{\partial}{\partial x_i} \right)$$

for some chart $\varphi \colon U \to M$. One can show that the maps of this form define a differentiable structure of dimension $2n$ in $TM$.

Now we are ready to introduce the notion of a vector field.

**Definition 1.82.** A *vector field* on a manifold $M$ is a map $X \colon M \to TM$ such that $X(x) \in T_x M$ for each $x \in M$.

Using the notion of differentiability between manifolds, one can also speak of a differentiable vector field.

**1.6.3. Differential equations.** Let $M$ be a manifold. Given a continuous vector field $X \colon M \to TM$, consider the ordinary differential equation

$$x' = X(x). \tag{1.103}$$

More generally, one can consider time-dependent vector fields.

**Definition 1.83.** A curve $x \colon (a,b) \to M$ of class $C^1$ is said to be a *solution* of equation (1.103) if $x'(t) = X(x(t))$ for every $t \in (a,b)$, where $x'(t)$ is the tangent vector to the curve $s \mapsto x(t+s)$ at $s = 0$.

Using charts one can establish the following manifold version of the Picard–Lindelöf theorem (Theorem 1.18).

**Theorem 1.84.** *If $X$ is a differentiable vector field on $M$, then for each $(t_0, x_0) \in \mathbb{R} \times M$ there exists a unique solution of the initial value problem*

$$\begin{cases} x' = X(x), \\ x(t_0) = x_0 \end{cases}$$

*in some open interval containing $t_0$.*

More generally, all local aspects of the theory of ordinary differential equations in $\mathbb{R}^n$ can be extended to arbitrary manifolds. In addition to the Picard–Lindelöf theorem, this includes all other results concerning the existence of solutions (such as Peano's theorem) and the dependence of solutions on the initial conditions (see Theorems 1.40 and 1.42).

On the other hand, some nonlocal aspects differ substantially in the case of differential equations on manifolds. In particular, the following result is a consequence of Theorem 1.46.

**Theorem 1.85.** *If $X$ is a differentiable vector field on a compact manifold, then all solutions of the equation $x' = X(x)$ are global, that is, all solutions are defined for $t \in \mathbb{R}$.*

In a similar manner to that in Proposition 1.13, this implies that for a differentiable vector field on a compact manifold the solutions of the equation $x' = X(x)$ give rise to a flow. More precisely, the following statement holds.

**Theorem 1.86.** *If $X$ is a differentiable vector field on a compact manifold $M$ and $x(t, x_0)$ is the solution of the initial value problem*

$$\begin{cases} x' = X(x), \\ x(0) = x_0, \end{cases}$$

*then the family of transformations $\varphi_t \colon M \to M$ defined by $\varphi_t(x_0) = x(t, x_0)$ is a flow.*

The following are examples of differential equations on manifolds.

**Example 1.87.** Let $M = S^1$. Given a vector field $X \colon M \to TM$, each vector $X(x)$ is either zero or is tangent to the circle. If $X$ does not take the value zero, then the whole $S^1$ is an orbit of the equation $x' = X(x)$. If $X$ has exactly $p$ zeros, then there are $2p$ orbits.

**Example 1.88.** Let $M = S^2 \subset \mathbb{R}^3$ and consider a vector field $X \colon M \to TM$ such that for $x \in S^2$ other than the North and South poles the vector $X(x)$ is nonzero and horizontal. Then the equation $x' = X(x)$ has critical points at the North and South poles and all other orbits are horizontal circles (obtained from intersecting the sphere with horizontal planes).

**Example 1.89.** Let $M = \mathbb{T}^2 = S^1 \times S^1$. Since $S^1$ can be seen as the interval $[0, 1]$ with the endpoints identified, the torus $\mathbb{T}^2$ can be seen as the square $[0, 1] \times [0, 1]$ with two points $x, y \in Q$ identified whenever $x - y \in \mathbb{Z}^2$. Thus, a differential equation in the torus can be seen as the equation

$$\begin{cases} x' = f(x, y), \\ y' = g(x, y) \end{cases} \tag{1.104}$$

in $\mathbb{R}^2$, for some functions $f, g \colon \mathbb{R}^2 \to \mathbb{R}$ such that

$$f(x + k, y + l) = f(x, y) \quad \text{and} \quad g(x + k, y + l) = g(x, y)$$

for every $x, y \in \mathbb{R}$ and $k, l \in \mathbb{Z}$.

For example, if $f = \alpha$ and $g = \beta$ for some constants $\alpha, \beta \in \mathbb{R}$, then the solutions of equation (1.104) give rise to the flow

$$\varphi_t(x, y) = (x + \alpha t, y + \beta t) \bmod 1.$$

If $\alpha = 0$, then the image of each solution is a "vertical" periodic orbit. If $\alpha \neq 0$ and $\beta/\alpha \in \mathbb{Q}$, then the image of each solution is a periodic orbit with slope $\beta/\alpha$. Finally, if $\alpha \neq 0$ and $\beta/\alpha \notin \mathbb{Q}$, then the image of each solution is dense in the torus.

A detailed study of the theory of differential equations on manifolds clearly falls outside the scope of the book. In particular, some topological obstructions influence the behavior of the solutions of a differential equation. For example, any vector field $X$ on $S^2$ has zeros and thus, the equation $x' = X(x)$ has critical points. We refer the reader to [**3, 11**] for detailed treatments.

## 1.7. Exercises

**Exercise 1.1.** For the equation $x' = f(x)$, show that if $\cos t$ is a solution, then $-\sin t$ is also a solution.

**Exercise 1.2.** For the equation $x'' = f(x)$:

    a) show that if $1/(1+t)$ is a solution, then $1/(1-t)$ is also a solution;

    b) find $f$ such that $1/(1+t)$ is a solution.

**Exercise 1.3.** Find an autonomous equation having $t^2/(1+t)$ as a solution.

**Exercise 1.4.** For the equation $(x, y)' = (y, x)$ in $\mathbb{R}^2$, writing

$$x(t) = r \cosh \theta(t) \quad \text{and} \quad y(t) = r \sinh \theta(t),$$

with $r \in \mathbb{R}$, show that $\theta(t) = t + c$ for some constant $c \in \mathbb{R}$.

**Exercise 1.5.** Let $f \colon \mathbb{R} \to \mathbb{R}$ be a Lipschitz function and let $g \colon \mathbb{R} \to \mathbb{R}$ be a continuous function. Show that for the equation

$$\begin{cases} x' = f(x), \\ y' = g(x)y \end{cases}$$

the initial value problem (1.13) has a unique solution.

**Exercise 1.6.** Let $f \colon (a, b) \to \mathbb{R} \setminus \{0\}$ be a continuous function.

    a) Show that the function $g \colon (a, b) \to \mathbb{R}$ defined by

$$g(x) = \int_{x_0}^{x} \frac{1}{f(y)} \, dy$$

    is invertible for each $x_0 \in \mathbb{R}$.

    b) Show that the initial value problem (1.13) has a unique solution. Hint: For a solution $x(t)$, compute the derivative of the function $t \mapsto g(x(t))$.

**Exercise 1.7.** Let $f\colon \mathbb{R}^n \to \mathbb{R}^n$ be a function of class $C^1$ and let $\langle \cdot, \cdot \rangle$ be an inner product in $\mathbb{R}^n$.

    a) Show that if $\langle f(x), x \rangle \le 0$ for every $x \in \mathbb{R}^n$, then each solution of the initial value problem (1.13) is defined for all $t > t_0$.

    b) Show that if $g\colon \mathbb{R}^n \to \mathbb{R}$ is a differentiable function such that

$$g(x) \ge \|x\|^2 \quad \text{and} \quad \langle f(x), \nabla g(x) \rangle < 0$$

for every $x \in \mathbb{R}^n \setminus \{0\}$ (where $\|x\|^2 = \langle x, x \rangle$ and $\nabla g$ is the gradient of $g$), then each solution of the initial value problem (1.13) is defined for all $t > t_0$.

**Exercise 1.8.** Write the equation

$$\begin{cases} x' = -ay, \\ y' = ax \end{cases}$$

in polar coordinates.

**Exercise 1.9.** Write the equation

$$\begin{cases} x' = \varepsilon x - y - x(x^2 + y^2), \\ y' = x + \varepsilon y - y(x^2 + y^2) \end{cases}$$

in polar coordinates.

**Exercise 1.10.** Let $u, v, w\colon [a, b] \to \mathbb{R}$ be continuous functions with $w > 0$ such that

$$u(t) \le v(t) + \int_a^t w(s)u(s)\, ds$$

for every $t \in [a, b]$. Show that

$$u(t) \le v(t) + \int_a^t w(s)v(s) \exp\left( \int_s^t w(u)\, du \right) ds$$

for every $t \in [a, b]$.

**Exercise 1.11.** Let $T\colon X \to X$ be a transformation of a complete metric space $X$ such that $T^m$ is a contraction for some $m \in \mathbb{N}$. Show that:

    a) $T$ has a unique fixed point $x_0 \in X$;

    b) for each $x \in X$ the sequence $T^n(x)$ converges to $x_0$ when $n \to \infty$.

**Exercise 1.12.** Show that the function $A \mapsto \|A\|$ defined by (1.40) is a norm in the space $M_n$ of $n \times n$ matrices with real entries (see Definition 1.26).

**Exercise 1.13.** Show that the norm $\|\cdot\|$ defined by (1.40) satisfies

$$\|AB\| \le \|A\| \cdot \|B\|$$

for every $A, B \in M_n$.

**Exercise 1.14.** Consider the set $Z \subset C(I)$ (see Proposition 1.30) of all Lipschitz functions $x \colon I \to \mathbb{R}^n$ in a given bounded set $I \subset \mathbb{R}^k$, with the same constant $L$ in (1.38), such that $x(0) = 0$ (when $0 \in I$). Show that $Z$ is a complete metric space with the distance

$$\bar{d}(x, y) = \sup \left\{ \frac{\|x(t) - y(t)\|}{\|t\|} : t \in I \setminus \{0\} \right\}.$$

**Exercise 1.15.** Let $\varphi_t$ be a flow defined by an equation $x' = f(x)$. Say how $f$ can be obtained from the flow.

**Exercise 1.16.** Show that if $f \colon \mathbb{R}^n \to \mathbb{R}^n$ is a bounded function of class $C^1$, then the equation $x' = f(x)$ defines a flow such that each function $\varphi_t$ is a homeomorphism (that is, a bijective continuous function with continuous inverse).

**Exercise 1.17.** Given $T > 0$, let $f \colon \mathbb{R} \times \mathbb{R}^n \to \mathbb{R}^n$ be a Lipschitz function such that

$$f(t, x) = f(t + T, x) \quad \text{for} \quad (t, x) \in \mathbb{R} \times \mathbb{R}^n.$$

Show that for each $(t_0, x_0) \in \mathbb{R} \times \mathbb{R}^n$ the solution $x(t, t_0, x_0)$ of the initial value problem (1.13) has maximal interval $\mathbb{R}$ and satisfies the identity

$$x(t, t_0, x_0) = x(t + T, t_0 + T, x_0) \quad \text{for} \quad t \in \mathbb{R}.$$

**Exercise 1.18.** Sketch the phase portrait of the equation $x'' = x^3 - x$, that is, of the equation $(x, y)' = (y, x^3 - x)$ in $\mathbb{R}^2$.

**Exercise 1.19.** Sketch the phase portrait of the equation:

a) $\begin{cases} x' = y(y^2 - x^2), \\ y' = -x(y^2 - x^2); \end{cases}$

b) $\begin{cases} x' = -y + x(1 - x^2 - y^2), \\ y' = x + y(1 - x^2 - y^2); \end{cases}$

c) $\begin{cases} x' = x(10 - x^2 - y^2), \\ y' = y(1 - x^2 - y^2). \end{cases}$

**Exercise 1.20.** Find all periodic orbits of each equation in Exercise 1.19.

**Exercise 1.21.** For each $\lambda \in \mathbb{R}$, sketch the phase portrait of the equation:

a) $x'' = \lambda x(x - 1)$;

b) $x'' = x(x - \lambda)$;

c) $\begin{cases} x' = -y + x(\lambda - x^2 - y^2), \\ y' = x + y(\lambda - x^2 - y^2). \end{cases}$

**Exercise 1.22.** Find an integral for the equation

$$\begin{cases} x' = x + 3y^2, \\ y' = -2x - y. \end{cases}$$

**Exercise 1.23.** Consider the equation in polar coordinates

$$\begin{cases} r' = 0, \\ \theta' = (r^2 - 1)(r^2 \cos^2 \theta + r \sin \theta + 1). \end{cases}$$

a) Sketch the phase portrait.

b) Find all global solutions.

c) Find whether the equation is conservative.

d) Show that in a neighborhood of the origin the periods $T(r)$ of the periodic orbits satisfy

$$T(r) = 2\pi + ar^2 + o(r^2),$$

for some constant $a \neq 0$ (given a function $g \colon \mathbb{R} \to \mathbb{R}$ such that $g(x)/x^k \to 0$ when $x \to 0$, we write $g(x) = o(x^k)$).

**Solutions.**

**1.2** b) $f(x) = 2x^3$.

**1.3** $[(x' + x)/(2 - x')]' = 1$, that is, $(x, y)' = (y, 2(y - 1)(y - 2)/(2 + x))$.

**1.8** $(r, \theta)' = (0, a)$.

**1.9** $(r, \theta)' = (\varepsilon r - r^3, 1)$.

**1.15** $f(x) = (\partial/\partial t)\varphi_t(x)|_{t=0}$.

**1.20** a) There are no periodic orbits.

b) $\{(x, y) \in \mathbb{R}^2 : x^2 + y^2 = 1\}$.

c) There are no periodic orbits.

**1.22** $E(x, y) = x^2 + xy + y^3$.

**1.23** b) All solutions are global.

c) It is conservative.

# Chapter 2

# Linear Equations and Conjugacies

This chapter is dedicated to the study of the qualitative properties of linear ordinary differential equations and their perturbations. We first introduce the notion of a fundamental solution, which contains complete information about the solutions of a linear equation. In particular, we describe the fundamental solutions of the equations with constant coefficients and periodic coefficients. We then give a complete description of the phase portraits of the equations with constant coefficients in dimension 1 and dimension 2. We also present the variation of parameters formula for the solutions of the perturbations of a linear equation. Finally, we introduce the notion of a conjugacy between the solutions of two linear differential equations with constant coefficients, and we discuss the characterization of the differentiable, topological and linear conjugacies in terms of the coefficients of the equations. For additional topics we refer the reader to [4, 15, 17].

## 2.1. Nonautonomous linear equations

In this section we consider equations in $\mathbb{R}^n$ of the form

$$x' = A(t)x, \qquad (2.1)$$

where the matrices $A(t) \in M_n$ vary continuously with $t \in \mathbb{R}$ (we recall that $M_n$ is the set of $n \times n$ matrices with real entries). Equation (2.1) is called a *linear equation*.

**2.1.1. Space of solutions.** One can easily verify that $f(t, x) = A(t)x$ is continuous and locally Lipschitz in $x$. It follows from the Picard–Lindelöf theorem (Theorem 1.18) that the initial value problem (1.13), that is,

$$\begin{cases} x' = A(t)x, \\ x(t_0) = x_0, \end{cases} \tag{2.2}$$

has a unique solution in some open interval containing $t_0$.

**Proposition 2.1.** *All solutions of equation (2.1) have maximal interval $\mathbb{R}$.*

**Proof.** We consider only times $t \geq t_0$. The argument for $t \leq t_0$ is entirely analogous. If $x = x(t)$ is a solution of equation (2.1), then

$$x(t) = x(t_0) + \int_{t_0}^{t} A(s)x(s)\,ds$$

for $t$ in some open interval containing $t_0$, and thus,

$$\|x(t)\| \leq \|x(t_0)\| + \int_{t_0}^{t} \|A(s)\| \cdot \|x(s)\|\,ds$$

for $t \geq t_0$ in that interval. It follows from Gronwall's lemma (Proposition 1.39) that

$$\|x(t)\| \leq \|x(t_0)\| \exp \int_{t_0}^{t} \|A(s)\|\,ds,$$

for the same values of $t$. Since the function $t \mapsto A(t)$ is continuous, the integral $\int_{t_0}^{t} \|A(s)\|\,ds$ is well defined and is finite for every $t_0$ and $t$. Thus, it follows from Theorem 1.46 that the maximal interval of the solution $x$ is $\mathbb{R}$. $\qquad\square$

Using the notion of global solution in Definition 1.56, one can rephrase Proposition 2.1 as follows: all solutions of equation (2.1) are global.

The following result shows the set of all solutions of equation (2.1) is a linear space.

**Proposition 2.2.** *If $x_1, x_2 \colon \mathbb{R} \to \mathbb{R}^n$ are solutions of equation (2.1), then $c_1 x_1 + c_2 x_2$ is a solution of equation (2.1) for each $c_1, c_2 \in \mathbb{R}$.*

**Proof.** We have

$$\begin{aligned}
(c_1 x_1 + c_2 x_2)' &= c_1 x_1' + c_2 x_2' \\
&= c_1 A(t) x_1 + c_2 A(t) x_2 \\
&= A(t)(c_1 x_1 + c_2 x_2),
\end{aligned}$$

and thus, $c_1 x_1 + c_2 x_2$ is a solution of equation (2.1). $\qquad\square$

**Example 2.3.** Consider the equation

$$\begin{pmatrix} x \\ y \end{pmatrix}' = \begin{pmatrix} 0 & 1 \\ -1 & 0 \end{pmatrix} \begin{pmatrix} x \\ y \end{pmatrix}, \tag{2.3}$$

whose solutions are given by (1.7), with $r \geq 0$ and $c \in \mathbb{R}$. We have

$$\begin{pmatrix} x(t) \\ y(t) \end{pmatrix} = \begin{pmatrix} r \cos c \cos t + r \sin c \sin t \\ -r \cos c \sin t + r \sin c \cos t \end{pmatrix}$$

$$= r \cos c \begin{pmatrix} \cos t \\ -\sin t \end{pmatrix} + r \sin c \begin{pmatrix} \sin t \\ \cos t \end{pmatrix}.$$

This shows that the set of all solutions of equation (2.3) is a linear space of dimension 2, generated by the (linearly independent) vectors

$$(\cos t, -\sin t) \quad \text{and} \quad (\sin t, \cos t).$$

More generally, we have the following result.

**Proposition 2.4.** *The set of all solutions of equation* (2.1) *is a linear space of dimension $n$.*

**Proof.** Let $e_1, \ldots, e_n$ be a basis of $\mathbb{R}^n$. For $i = 1, \ldots, n$, let $x_i = x_i(t)$ be the solution of the initial value problem (2.2) with $x_0 = e_i$. For an arbitrary $x_0 \in \mathbb{R}^n$, the solution of problem (2.2) can be obtained as follows. Writing $x_0 = \sum_{i=1}^n c_i e_i$, one can easily verify that the function

$$x(t) = \sum_{i=1}^n c_i x_i(t)$$

is the solution of equation (2.1) with $x(t_0) = x_0$. In particular, the space of solutions of equation (2.1) is generated by the functions $x_1(t), \ldots, x_n(t)$. In order to show that these are linearly independent, let us assume that

$$\sum_{i=1}^n c_i x_i(t) = 0$$

for some constants $c_1, \ldots, c_n \in \mathbb{R}$ and every $t \in \mathbb{R}$. Taking $t = t_0$, we obtain $\sum_{i=1}^n c_i e_i = 0$, and hence $c_1 = c_2 = \cdots = c_n = 0$ (because $e_1, \ldots, e_n$ is a basis). Thus, the functions $x_1(t), \ldots, x_n(t)$ are linearly independent. $\square$

**2.1.2. Fundamental solutions.** In order to describe all solutions of equation (2.1) it is useful to introduce the following notion.

**Definition 2.5.** A function $X(t)$ with values in $M_n$ whose columns form a basis of the space of solutions of equation (2.1) is called a *fundamental solution* of the equation.

We note that any fundamental solution is a function $X \colon \mathbb{R} \to M_n$ satisfying

$$X'(t) = A(t)X(t) \tag{2.4}$$

for every $t \in \mathbb{R}$, with the derivative of $X(t)$ computed entry by entry.

Now let $X(t)$ be a fundamental solution of equation (2.1). For each vector $c \in \mathbb{R}^n$, the product $X(t)c$ is a linear combination of the columns of $X(t)$. More precisely, if $x_1(t), \ldots, x_n(t)$ are the columns of $X(t)$ and $c = (c_1, \ldots, c_n)$, then

$$X(t)c = \sum_{i=1}^{n} c_i x_i(t).$$

This shows that the solutions of equation (2.1) are exactly the functions of the form $X(t)c$ with $c \in \mathbb{R}^n$, where $X(t)$ is any fundamental solution.

**Example 2.6.** Let $f \colon \mathbb{R}^n \to \mathbb{R}^n$ be a function of class $C^1$ such that the equation $x' = f(x)$ has only global solutions. Given a solution $\varphi_t(x)$ with $\varphi_0(x) = x$, consider the matrices

$$A(t) = d_{\varphi_t(x)} f.$$

The equation

$$y' = A(t)y$$

is called the *linear variational equation* of $x' = f(x)$ along the solution $\varphi_t(x)$. It follows from the identity

$$\frac{\partial}{\partial t} \varphi_t(x) = f(\varphi_t(x))$$

that if the map $(t, x) \mapsto \varphi_t(x)$ is sufficiently regular, then

$$\frac{\partial}{\partial t} d_x \varphi_t = d_{\varphi_t(x)} f \, d_x \varphi_t = A(t) d_x \varphi_t.$$

This shows that the function $X(t) = d_x \varphi_t$ satisfies the identity

$$X'(t) = A(t)X(t),$$

and thus also

$$(X(t)c)' = A(t)(X(t)c)$$

for each $c \in \mathbb{R}^n$. Since

$$X(0) = d_x \varphi_0 = d_x \mathrm{Id} = \mathrm{Id},$$

the columns of $X(t)$ are linearly independent, and hence $X(t)$ is a fundamental solution of the linear variational equation of $x' = f(x)$ along the solution $\varphi_t(x)$.

Now we establish an auxiliary result.

**Proposition 2.7.** *If $X(t)$ is a fundamental solution of equation (2.1), then its columns are linearly independent for every $t \in \mathbb{R}$.*

**Proof.** Let $x_1(t), \ldots, x_n(t)$ be the columns of a fundamental solution $X(t)$. Given $t_0 \in \mathbb{R}$, we show that the vectors $x_1(t_0), \ldots, x_n(t_0)$ are linearly independent. Assume that

$$\sum_{i=1}^{n} c_i x_i(t_0) = 0$$

for some constants $c_1, \ldots, c_n \in \mathbb{R}$. Then the solution

$$x(t) = \sum_{i=1}^{n} c_i x_i(t) \tag{2.5}$$

of equation (2.1) satisfies $x(t_0) = 0$, and it follows from the uniqueness of solutions that $x(t) = 0$ for every $t \in \mathbb{R}$. Since the functions $x_1(t), \ldots, x_n(t)$ form a basis of the space of solutions (because $X(t)$ is a fundamental solution), it follows from (2.5) that $c_1 = \cdots = c_n = 0$. Thus, the vectors $x_1(t_0), \ldots, x_n(t_0)$ are linearly independent. $\square$

Proposition 2.7 has the following consequence.

**Proposition 2.8.** *The solution of the initial value problem* (2.2) *is given by*

$$x(t) = X(t)X(t_0)^{-1} x_0, \quad t \in \mathbb{R}, \tag{2.6}$$

*where $X(t)$ is any fundamental solution of equation* (2.1).

**Proof.** It follows from Proposition 2.7 that the matrix $X(t_0)$ is invertible, and thus, the function $x(t)$ in (2.6) is well defined. Moreover,

$$x(t_0) = X(t_0)X(t_0)^{-1} x_0 = x_0,$$

and it follows from identity (2.4) that

$$\begin{aligned} x'(t) &= X'(t)X(t_0)^{-1} x_0 \\ &= A(t)X(t)X(t_0)^{-1} x_0 = A(t)x(t). \end{aligned}$$

This shows that $x(t)$ is the solution of the initial value problem (2.2). $\square$

Any two fundamental solutions are related as follows.

**Proposition 2.9.** *If $X(t)$ and $Y(t)$ are fundamental solutions of equation* (2.1), *then there exists an invertible matrix $C \in M_n$ such that $Y(t) = X(t)C$ for every $t \in \mathbb{R}$.*

**Proof.** Since $X(t)$ and $Y(t)$ are fundamental solutions, each column of $Y(t)$ is a linear combination of the columns of $X(t)$ and vice-versa. Hence, there exist matrices $C, D \in M_n$ such that

$$Y(t) = X(t)C \quad \text{and} \quad X(t) = Y(t)D$$

for every $t \in \mathbb{R}$. Thus,

$$Y(t) = X(t)C = Y(t)DC. \qquad (2.7)$$

On the other hand, by Proposition 2.7, the columns of the matrix $Y(t)$ are linearly independent for each $t \in \mathbb{R}$. Therefore, $Y(t)$ is invertible and it follows from (2.7) that $DC = \mathrm{Id}$. In particular, the matrix $C$ is invertible. $\qquad\square$

It also follows from Proposition 2.7 that if $X(t)$ is a fundamental solution of a linear equation, then $\det X(t) \neq 0$ for every $t \in \mathbb{R}$. In fact, we have a formula for the determinant.

**Theorem 2.10** (Liouville's formula). *If $X(t)$ is a fundamental solution of equation (2.1), then*

$$\det X(t) = \det X(t_0) \exp \int_{t_0}^{t} \mathrm{tr}\, A(s)\, ds$$

*for every $t, t_0 \in \mathbb{R}$.*

**Proof.** One can easily verify that

$$\frac{d}{dt} \det X(t) = \sum_{i=1}^{n} \det D_i(t), \qquad (2.8)$$

where $D_i(t)$ is the $n \times n$ matrix obtained from $X(t)$ taking the derivative of the $i$th line, and leaving all other lines unchanged. Writing $X(t) = (x_{ij}(t))$ and $A(t) = (a_{ij}(t))$, it follows from identity (2.4) that

$$x'_{ij}(t) = \sum_{k=1}^{n} a_{ik}(t)x_{kj}(t)$$

for every $t \in \mathbb{R}$ and $i, j = 1, \ldots, n$. In particular, the $i$th line of $D_i(t)$ is

$$(x'_{i1}(t), \ldots, x'_{in}(t)) = \left( \sum_{k=1}^{n} a_{ik}(t)x_{k1}(t), \ldots, \sum_{k=1}^{n} a_{ik}(t)x_{kn}(t) \right)$$

$$= \sum_{k=1}^{n} a_{ik}(t)(x_{k1}(t), \ldots, x_{kn}(t)).$$

Since the last sum is a linear combination of the lines of $X(t)$, it follows from (2.8) and the properties of the determinant that

$$\frac{d}{dt} \det X(t) = \sum_{i=1}^{n} a_{ii}(t) \det X(t)$$

$$= \mathrm{tr}\, A(t) \det X(t).$$

In other words, the function $u(t) = \det X(t)$ is a solution of the equation $u' = \operatorname{tr} A(t)u$. Therefore,

$$u(t) = u(t_0) \exp \int_{t_0}^{t} \operatorname{tr} A(s) \, ds,$$

which yields the desired result. □

## 2.2. Equations with constant coefficients

In this section we consider the particular case of the linear equations in $\mathbb{R}^n$ of the form

$$x' = Ax, \tag{2.9}$$

where $A$ is an $n \times n$ matrix with real entries. We note that the function $f(t, x) = Ax$ is of class $C^1$, and thus, it is continuous and locally Lipschitz in $x$.

**2.2.1. Exponential of a matrix.** In order to solve equation (2.9), we first recall the notion of the exponential of a matrix.

**Definition 2.11.** We define the *exponential* of a matrix $A \in M_n$ by

$$e^A = \sum_{k=0}^{\infty} \frac{1}{k!} A^k, \tag{2.10}$$

with the convention that $e^0 = \operatorname{Id}$.

**Proposition 2.12.** *The power series in* (2.10) *is convergent for every matrix $A$.*

**Proof.** One can easily verify that $\|A^k\| \leq \|A\|^k$ for each $k \in \mathbb{N}$, and thus,

$$\sum_{k=0}^{\infty} \frac{1}{k!} \|A^k\| \leq \sum_{k=0}^{\infty} \frac{1}{k!} \|A\|^k = e^{\|A\|} < +\infty. \tag{2.11}$$

In other words, the series in (2.10) is absolutely convergent, and thus it is also convergent. □

**Example 2.13.** If $A = \operatorname{Id}$, then $A^k = \operatorname{Id}$ for each $k \in \mathbb{N}$, and thus,

$$e^{\operatorname{Id}} = \sum_{k=0}^{\infty} \frac{1}{k!} \operatorname{Id} = e \operatorname{Id}.$$

**Example 2.14.** If $A$ is an $n \times n$ diagonal matrix with entries $\lambda_1, \ldots, \lambda_n$ in the diagonal, then

$$e^A = \sum_{k=0}^{\infty} \frac{1}{k!} A^k = \sum_{k=0}^{\infty} \frac{1}{k!} \begin{pmatrix} \lambda_1^k & & 0 \\ & \ddots & \\ 0 & & \lambda_n^k \end{pmatrix}$$

$$= \begin{pmatrix} \sum_{k=0}^{\infty} \frac{1}{k!} \lambda_1^k & & 0 \\ & \ddots & \\ 0 & & \sum_{k=0}^{\infty} \frac{1}{k!} \lambda_n^k \end{pmatrix}$$

$$= \begin{pmatrix} e^{\lambda_1} & & 0 \\ & \ddots & \\ 0 & & e^{\lambda_n} \end{pmatrix}.$$

**Example 2.15.** Consider the matrix

$$A = \begin{pmatrix} 0 & \mathrm{Id} \\ -\mathrm{Id} & 0 \end{pmatrix},$$

where $\mathrm{Id} \in M_n$ is the identity. We have

$$A^0 = \mathrm{Id}, \quad A^1 = A, \quad A^2 = -\mathrm{Id}, \quad A^3 = -A, \quad A^4 = \mathrm{Id},$$

and thus $A^{k+4} = A^k$ for each $k \in \mathbb{N}$. Hence,

$$e^{At} = \left( \frac{t^0}{0!} + \frac{t^4}{4!} + \cdots \right) \mathrm{Id} + \left( \frac{t^1}{1!} + \frac{t^5}{5!} + \cdots \right) A$$

$$- \left( \frac{t^2}{2!} + \frac{t^6}{6!} + \cdots \right) \mathrm{Id} - \left( \frac{t^3}{3!} + \frac{t^7}{7!} + \cdots \right) A$$

$$= \frac{e^{it} + e^{-it}}{2} \mathrm{Id} + \frac{e^{it} - e^{-it}}{2i} A$$

$$= \cos t \mathrm{Id} + \sin t A.$$

In order to describe a somewhat expedited method to compute the exponential of a matrix, we first recall a basic result from linear algebra.

**Theorem 2.16** (Jordan canonical form). *For each $A \in M_n$, there exists an invertible $n \times n$ matrix $S$ with entries in $\mathbb{C}$ such that*

$$S^{-1} A S = \begin{pmatrix} R_1 & & 0 \\ & \ddots & \\ 0 & & R_k \end{pmatrix}, \tag{2.12}$$

*where each block $R_j$ is an $n_j \times n_j$ matrix, for some $n_j \leq n$, of the form*

$$R_j = \begin{pmatrix} \lambda_j & 1 & & 0 \\ & \ddots & \ddots & \\ & & \ddots & 1 \\ 0 & & & \lambda_j \end{pmatrix}, \tag{2.13}$$

*where $\lambda_j$ is an eigenvalue of $A$.*

Using the Jordan canonical form in Theorem 2.16, the exponential of a matrix can be computed as follows.

**Proposition 2.17.** *If $A$ is a square matrix with the Jordan canonical form in (2.12), then*

$$e^A = S e^{S^{-1}AS} S^{-1} = S \begin{pmatrix} e^{R_1} & & 0 \\ & \ddots & \\ 0 & & e^{R_k} \end{pmatrix} S^{-1}. \tag{2.14}$$

**Proof.** We have $(S^{-1}AS)^k = S^{-1}A^k S$ for each $k \in \mathbb{N}$. Therefore,

$$e^{S^{-1}AS} = \sum_{k=0}^{\infty} \frac{1}{k!} (S^{-1}AS)^k$$

$$= S^{-1} \left( \sum_{k=0}^{\infty} \frac{1}{k!} A^k \right) S = S^{-1} e^A S.$$

On the other hand, one can easily verify that the exponential of a matrix in block form is obtained computing separately the exponential of each block, that is,

$$e^{S^{-1}AS} = \begin{pmatrix} e^{R_1} & & 0 \\ & \ddots & \\ 0 & & e^{R_k} \end{pmatrix}.$$

This yields the desired result. $\square$

When $A$ consists of a single block $R_j$ as in (2.13), the exponential $e^A$ can be computed explicitly as follows.

**Proposition 2.18.** *If a matrix $A \in M_n$ has the form $A = \lambda \mathrm{Id} + N$, where*

$$N = \begin{pmatrix} 0 & 1 & & 0 \\ & \ddots & \ddots & \\ & & \ddots & 1 \\ 0 & & & 0 \end{pmatrix}, \tag{2.15}$$

LIVERPOOL JOHN MOORES UNIVERSITY
LEARNING SERVICES

*then*

$$e^A = e^\lambda \left( \mathrm{Id} + N + \frac{1}{2!}N^2 + \cdots + \frac{1}{(n-1)!}N^{n-1} \right). \qquad (2.16)$$

**Proof.** Since the matrices $\lambda\mathrm{Id}$ and $N$ commute, one can show that

$$e^A = e^{\lambda\mathrm{Id}} e^N$$

(see Exercise 2.4). By Example 2.14, we have $e^{\lambda\mathrm{Id}} = e^\lambda\mathrm{Id}$. On the other hand, since $N^n = 0$, we obtain

$$e^N = \sum_{k=0}^{\infty} \frac{1}{k!}N^k = \sum_{k=0}^{n-1} \frac{1}{k!}N^k,$$

and thus,

$$e^A = e^\lambda\mathrm{Id} \sum_{k=0}^{n-1} \frac{1}{k!}N^k = e^\lambda \sum_{k=0}^{n-1} \frac{1}{k!}N^k.$$

This yields identity (2.16). $\qquad\qquad\square$

**2.2.2. Solving the equations.** In this section we describe the relation between a linear equation $x' = Ax$ and the exponential of a matrix. We start with an auxiliary result.

**Proposition 2.19.** *We have $(e^{At})' = Ae^{At}$ for each $t \in \mathbb{R}$, with the derivative of $e^{At}$ computed entry by entry.*

**Proof.** We have

$$e^{At} = \sum_{k=0}^{\infty} \frac{1}{k!}t^k A^k.$$

Since

$$\sum_{k=0}^{\infty} \frac{1}{k!}\|t^k A^k\| \le \sum_{k=0}^{\infty} \frac{1}{k!}\|tA\|^k = e^{\|tA\|} < +\infty,$$

each entry of the matrix $e^{At}$ is a power series in $t$ with radius of convergence $+\infty$. Hence, one can differentiate the series term by term to obtain

$$(e^{At})' = \sum_{k=1}^{\infty} \frac{1}{k!}kt^{k-1}A^k$$

$$= A\sum_{k=1}^{\infty} \frac{1}{(k-1)!}t^{k-1}A^{k-1}$$

$$= Ae^{At}$$

for each $t \in \mathbb{R}$. $\qquad\qquad\square$

Now we find all solutions of equation (2.9).

**Proposition 2.20.** *Given $(t_0, x_0) \in \mathbb{R} \times \mathbb{R}^n$, the solution of the equation $x' = Ax$ with initial condition $x(t_0) = x_0$ is given by*

$$x(t) = e^{A(t-t_0)} x_0, \quad t \in \mathbb{R}. \tag{2.17}$$

**Proof.** By the Picard–Lindelöf theorem (Theorem 1.18), the desired solution is unique, and by Proposition 2.1 its maximal interval is $\mathbb{R}$. To complete the proof, it is sufficient to observe that for the function $x(t)$ in (2.17) we have

$$x(t_0) = e^{A0} x_0 = e^0 x_0 = \mathrm{Id} x_0 = x_0,$$

and by Proposition 2.19,

$$x'(t) = A e^{A(t-t_0)} x_0 = A x(t).$$

This shows that $x(t)$ is the desired solution.  $\square$

### 2.2.3. Phase portraits.

In this section we describe the phase portraits of all linear equations with constant coefficients in $\mathbb{R}$ and in $\mathbb{R}^2$. The following example gives a complete description for scalar equations.

**Example 2.21.** Consider the scalar equation $x' = ax$, where $a \in \mathbb{R}$. Its phase portrait, shown in Figure 2.1, depends only on the sign of the constant $a$. Namely, when $a = 0$ there are only critical points. On the other hand, when $a \neq 0$ the origin is the only critical point, and the remaining orbits $\mathbb{R}^+$ and $\mathbb{R}^-$ have the direction indicated in Figure 2.1.

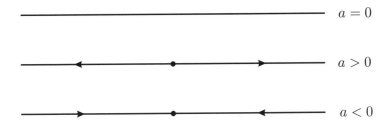

**Figure 2.1.** Phase portrait of the equation $x' = ax$.

Now we consider linear equations in $\mathbb{R}^2$ and we describe their phase portraits based on the Jordan canonical form.

**Example 2.22.** We first consider a matrix $A \in M_2$ with Jordan canonical form

$$\begin{pmatrix} \lambda_1 & 0 \\ 0 & \lambda_2 \end{pmatrix}$$

for some $\lambda_1, \lambda_2 \in \mathbb{R}$ (we emphasize that we are not assuming that $A$ is of this form, but only that its Jordan canonical form is of this form). Let

$v_1, v_2 \in \mathbb{R}^2$ be eigenvectors of $A$ associated respectively to the eigenvalues $\lambda_1$ and $\lambda_2$. The solutions of the equation $x' = Ax$ can be written in the form

$$x(t) = c_1 e^{\lambda_1 t} v_1 + c_2 e^{\lambda_2 t} v_2, \quad t \in \mathbb{R}, \tag{2.18}$$

with $c_1, c_2 \in \mathbb{R}$. Now we consider several cases.

**Case $\lambda_1 < \lambda_2 < 0$.** For $c_2 \neq 0$, we have

$$\frac{x(t)}{c_2 e^{\lambda_2 t}} \to v_2 \tag{2.19}$$

when $t \to +\infty$, and hence, the solution $x(t)$ has asymptotically the direction of $v_2$ (when $t \to +\infty$). This yields the phase portrait shown in Figure 2.2, and the origin is called a *stable node*.

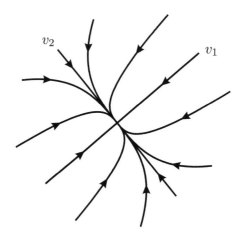

**Figure 2.2.** Case $\lambda_1 < \lambda_2 < 0$.

**Case $\lambda_1 > \lambda_2 > 0$.** Analogously, for $c_2 \neq 0$ property (2.19) holds when $t \to -\infty$, and hence, the solution $x(t)$ has asymptotically the direction of $v_2$ (when $t \to -\infty$). This yields the phase portrait shown in Figure 2.3, and the origin is called an *unstable node*.

**Case $\lambda_1 < 0 < \lambda_2$.** Now there is expansion along the direction of $v_2$ and contraction along the direction of $v_1$. The phase portrait is the one shown in Figure 2.4, and the origin is called a *saddle point*.

**Case $\lambda_1 = 0$ and $\lambda_2 > 0$.** The solution in (2.18) can now be written in the form

$$x(t) = c_1 v_1 + c_2 e^{\lambda_2 t} v_2, \quad t \in \mathbb{R}, \tag{2.20}$$

with $c_1, c_2 \in \mathbb{R}$. Thus, for $c_2 \neq 0$ the solution traverses a ray with the direction of $v_2$. On the other hand, the straight line passing through the

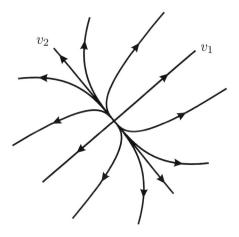

**Figure 2.3.** Case $\lambda_1 > \lambda_2 > 0$.

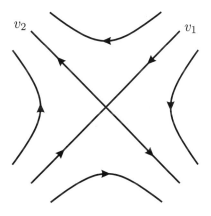

**Figure 2.4.** Case $\lambda_1 < 0 < \lambda_2$.

origin with the direction of $v_1$ consists entirely of critical points. This yields the phase portrait shown in Figure 2.5.

**Case $\lambda_1 = 0$ and $\lambda_2 < 0$.** The solution is again given by (2.20). Analogously, the phase portrait is the one shown in Figure 2.6.

**Case $\lambda_1 = \lambda_2 = \lambda < 0$.** The solution in (2.18) can now be written in the form

$$x(t) = e^{\lambda t} x_0, \quad t \in \mathbb{R},$$

with $x_0 \in \mathbb{R}^2$. We note that there is no privileged direction, unlike what happens in Figure 2.2, where there is an asymptote. Thus, the phase portrait is the one shown in Figure 2.7, and the origin is called a *stable node*.

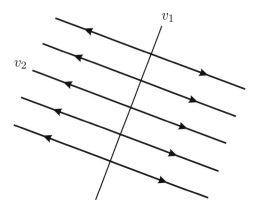

**Figure 2.5.** Case $\lambda_1 = 0$ and $\lambda_2 > 0$.

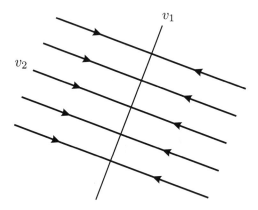

**Figure 2.6.** Case $\lambda_1 = 0$ and $\lambda_2 < 0$.

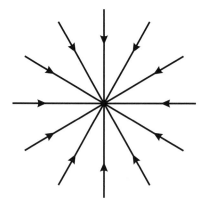

**Figure 2.7.** Case $\lambda_1 = \lambda_2 < 0$.

**Case** $\lambda_1 = \lambda_2 > 0$. Analogously, the phase portrait is the one shown in Figure 2.8, and the origin is called an *unstable node*.

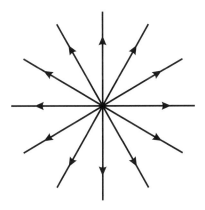

**Figure 2.8.** Case $\lambda_1 = \lambda_2 > 0$.

**Case** $\lambda_1 = \lambda_2 = 0$. In this case there are only critical points.

Now we consider the case of nondiagonal Jordan canonical forms with real eigenvalues.

**Example 2.23.** Consider a matrix $A \in M_2$ with Jordan canonical form

$$\begin{pmatrix} \lambda & 1 \\ 0 & \lambda \end{pmatrix}$$

for some $\lambda \in \mathbb{R}$. One can easily verify that the solutions of the equation $x' = Ax$ can be written in the form

$$x(t) = [(c_1 + c_2 t)v_1 + c_2 v_2]e^{\lambda t}, \quad t \in \mathbb{R}, \tag{2.21}$$

with $c_1, c_2 \in \mathbb{R}$, where $v_1, v_2 \in \mathbb{R}^2 \setminus \{0\}$ are vectors such that

$$Av_1 = \lambda v_1 \quad \text{and} \quad Av_2 = \lambda v_1 + v_2. \tag{2.22}$$

Whenever $c_1 \neq 0$ or $c_2 \neq 0$, we have

$$\frac{x(t)}{(c_1 + c_2 t)e^{\lambda t}} \to v_1$$

when $t \to +\infty$ and when $t \to -\infty$. Now we consider three cases.

**Case** $\lambda < 0$. The phase portrait is the one shown in Figure 2.9, and the origin is called a *stable node*.

**Case** $\lambda > 0$. The phase portrait is the one shown in Figure 2.10, and the origin is called an *unstable node*.

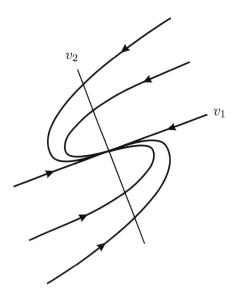

**Figure 2.9.** Case $\lambda < 0$.

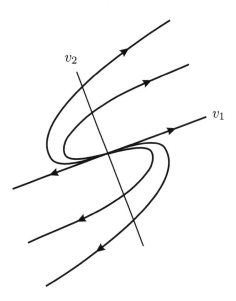

**Figure 2.10.** Case $\lambda > 0$.

**Case** $\lambda = 0$. The solution in (2.21) can now be written in the form

$$x(t) = (c_1 v_1 + c_2 v_2) + c_2 t v_1, \quad t \in \mathbb{R},$$

with $c_1, c_2 \in \mathbb{R}$, where $v_1, v_2 \in \mathbb{R}^2 \setminus \{0\}$ are vectors satisfying (2.22). In particular, the straight line passing through the origin with the direction of

$v_1$ consists entirely of critical points. The phase portrait is the one shown in Figure 2.11.

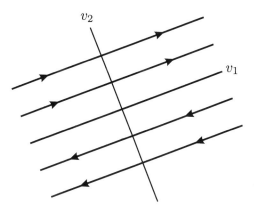

**Figure 2.11.** Case $\lambda = 0$.

Finally, we consider the case of nonreal eigenvalues. We recall that for matrices with real entries the nonreal eigenvalues always occur in pairs of conjugate complex numbers. Moreover, the eigenvectors associated to these eigenvalues are necessarily in $\mathbb{C}^2 \setminus \mathbb{R}^2$, and thus, it is natural to use complex variables.

**Example 2.24.** Let $A \in M_2$ be a $2 \times 2$ matrix with eigenvalues $a + ib$ and $a - ib$, for some $b \neq 0$. The solutions of the equation $x' = Ax$ in $\mathbb{C}^2$ are given by

$$x(t) = c_1 e^{(a+ib)t} v_1 + c_2 e^{(a-ib)t} v_2, \quad t \in \mathbb{R},$$

with $c_1, c_2 \in \mathbb{C}$, where $v_1, v_2 \in \mathbb{C}^2 \setminus \{0\}$ are eigenvectors associated respectively to $a + ib$ and $a - ib$. Since the matrix $A$ has real entries, one can take $v_2 = \overline{v_1}$. Indeed, it follows from $Av_1 = (a + ib)v_1$ that

$$A\overline{v_1} = \overline{Av_1} = \overline{(a + ib)v_1} = (a - ib)\overline{v_1},$$

and thus, $\overline{v_1}$ is an eigenvector associated to the eigenvalue $a - ib$. Hence, taking $c_2 = \overline{c_1}$, we obtain

$$\begin{aligned}
x(t) &= e^{at}\big[c_1 \cos(bt) + c_1 i \sin(bt)\big]v_1 + e^{at}\big[\overline{c_1}\cos(bt) - \overline{c_1}i\sin(bt)\big]\overline{v_1} \\
&= 2e^{at}\cos(bt)\operatorname{Re}(c_1 v_1) - 2e^{at}\sin(bt)\operatorname{Im}(c_1 v_1).
\end{aligned} \quad (2.23)$$

We note that the vectors

$$u_1 = 2\operatorname{Re}(c_1 v_1) \quad \text{and} \quad u_2 = -2\operatorname{Im}(c_1 v_1)$$

are in $\mathbb{R}^2$. Now we consider three cases.

**Case** $a = 0$. In this case the solution in (2.23) can be written in the form

$$x(t) = \cos(bt)u_1 + \sin(bt)u_2.$$

Thus, the phase portrait is the one shown in Figure 2.12 or the one obtained from it by reversing the direction of motion (this corresponds to change the sign of $b$). The origin is called a *center*.

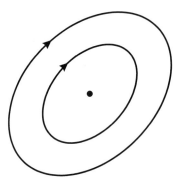

**Figure 2.12.** Case $a = 0$.

**Case** $a > 0$. The phase portrait is the one shown in Figure 2.13, and the origin is called a *stable focus*.

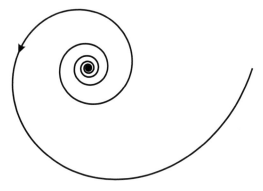

**Figure 2.13.** Case $a > 0$.

**Case** $a < 0$. The phase portrait is the one shown in Figure 2.14, and the origin is called an *unstable focus*.

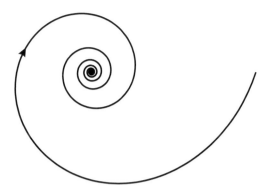

**Figure 2.14.** Case $a < 0$.

## 2.3. Variation of parameters formula

In this section we present a result giving the solutions of a class of perturbations of the linear equation (2.1). We continue to assume that the matrices $A(t) \in M_n$ vary continuously with $t \in \mathbb{R}$.

**Theorem 2.25** (Variation of parameters formula). *Let $b \colon \mathbb{R} \to \mathbb{R}^n$ be a continuous function. Given $(t_0, x_0) \in \mathbb{R} \times \mathbb{R}^n$, the solution of the initial value problem*

$$\begin{cases} x' = A(t)x + b(t), \\ x(t_0) = x_0 \end{cases} \tag{2.24}$$

*has maximal interval $\mathbb{R}$ and is given by*

$$x(t) = X(t)X(t_0)^{-1}x_0 + \int_{t_0}^{t} X(t)X(s)^{-1}b(s)\,ds, \tag{2.25}$$

*where $X(t)$ is any fundamental solution of equation (2.1), with the integral computed component by component.*

**Proof.** We first observe that the function

$$f(t, x) = A(t)x + b(t)$$

is continuous and locally Lipschitz in $x$. Thus, by the Picard–Lindelöf theorem (Theorem 1.18), the initial value problem (2.24) has a unique solution. Moreover, since the function $s \mapsto X(t)X(s)^{-1}b(s)$ is continuous (because the entries of the inverse of a matrix $B$ are continuous functions of the entries of $B$), the integral in (2.25) is well defined and is finite for every $t_0$ and $t$. Thus, it follows from Theorem 1.46 that the function $x(t)$ is defined for all

$t \in \mathbb{R}$. Moreover, $x(t_0) = x_0$, and using (2.4) we obtain

$$x'(t) = X'(t)X(t_0)^{-1}x_0 + \int_{t_0}^t X'(t)X(s)^{-1}b(s)\,ds + X(t)X(t)^{-1}b(t)$$

$$= A(t)X(t)X(t_0)^{-1}x_0 + A(t)\int_{t_0}^t X(t)X(s)^{-1}b(s)\,ds + b(t)$$

$$= A(t)x(t) + b(t).$$

This yields the desired result. □

The equations

$$x' = A(t)x \quad \text{and} \quad x' = A(t)x + b(t)$$

are often referred to, respectively, as *homogeneous equation* and *nonhomogeneous equation*.

We also obtain a Variation of parameters formula in the particular case of perturbations of the autonomous linear equation (2.9).

**Theorem 2.26** (Variation of parameters formula). *Let* $b \colon \mathbb{R} \to \mathbb{R}^n$ *be a continuous function. Given* $(t_0, x_0) \in \mathbb{R} \times \mathbb{R}^n$, *the solution of the initial value problem*

$$\begin{cases} x' = Ax + b(t), \\ x(t_0) = x_0 \end{cases} \tag{2.26}$$

*has maximal interval* $\mathbb{R}$ *and is given by*

$$x(t) = e^{A(t-t_0)}x_0 + \int_{t_0}^t e^{A(t-s)}b(s)\,ds. \tag{2.27}$$

**Proof.** By Theorem 2.25, the initial value problem (2.26) has a unique solution, with maximal interval $\mathbb{R}$. Hence, it is sufficient to observe that for the function $x(t)$ in (2.27) we have

$$x(t_0) = e^{A0}x_0 = \mathrm{Id}x_0 = x_0,$$

as well as

$$x'(t) = Ae^{A(t-t_0)}x_0 + \int_{t_0}^t Ae^{A(t-s)}b(s)\,ds + e^{A(t-t)}b(t)$$

$$= A\left(e^{A(t-t_0)}x_0 + \int_{t_0}^t e^{A(t-s)}b(s)\,ds\right) + b(t)$$

$$= Ax(t) + b(t).$$

This completes the proof of the theorem. □

We note that Theorem 2.26 is not an immediate consequence of Theorem 2.25. This is due to the fact that if $X(t)$ is a fundamental solution of equation (2.1), then in general $X(t)X(s)^{-1}$ may not be equal to $X(t-s)$, although this property holds for equations with constant coefficients (see Exercise 2.3).

Now we establish an auxiliary result.

**Proposition 2.27.** *If a matrix $A \in M_n$ has only eigenvalues with negative real part, then there exist constants $c, d > 0$ such that*

$$\|e^{At}\| \le ce^{-dt}, \quad t \ge 0, \tag{2.28}$$

*with the norm in (1.40) obtained from any given norm $\|\cdot\|$ in $\mathbb{R}^n$.*

**Proof.** Since the matrix $A$ has only eigenvalues with negative real part, it follows from (2.14) and (2.16) that for each entry $a_{ij}(t)$ of $e^{At}$ there exist constants $c_{ij}, d_{ij} > 0$ and a polynomial $p_{ij}$ such that

$$|a_{ij}(t)| \le c_{ij}e^{-d_{ij}t}|p_{ij}(t)|, \quad t \ge 0.$$

When $p_{ij}$ is not identically zero, we have

$$\limsup_{t \to +\infty} \frac{1}{t} \log \left( e^{-d_{ij}t}|p_{ij}(t)| \right) = -d_{ij},$$

and hence, given $\varepsilon > 0$, there exists $\bar{c}_{ij} > 0$ such that

$$e^{-d_{ij}t}|p_{ij}(t)| \le \bar{c}_{ij}e^{-(d_{ij}-\varepsilon)t}, \quad t \ge 0.$$

This yields the inequality

$$|a_{ij}(t)| \le c_{ij}\bar{c}_{ij}e^{-(d_{ij}-\varepsilon)t}, \quad t \ge 0.$$

Taking $\varepsilon$ sufficiently small so that

$$d := \min\left\{ d_{ij} - \varepsilon : i, j = 1, \ldots, n \right\} > 0,$$

we obtain

$$|a_{ij}(t)| \le \bar{c}e^{-dt}, \quad t \ge 0, \tag{2.29}$$

where

$$\bar{c} = \max\left\{ c_{ij}\bar{c}_{ij} : i, j = 1, \ldots, n \right\}.$$

Now we recall that all norms in $\mathbb{R}^m$ are equivalent, that is, given two norms $\|\cdot\|$ and $\|\cdot\|'$, there exists $C \ge 1$ such that

$$C^{-1}\|x\| \le \|x\|' \le C\|x\|$$

for every $x \in \mathbb{R}^m$. Since $M_n$ can be identified with $\mathbb{R}^{n^2}$, taking $m = n^2$ it follows from (2.29) that inequality (2.28) holds for some constant $c$. $\qquad\square$

The following is an application of Theorem 2.26 to a particular class of equations of the form $x' = Ax + b(t)$.

**Example 2.28.** Under the hypotheses of Theorem 2.26, let us assume that the matrix $A$ has only eigenvalues with negative real part and that the function $b$ is bounded. Using (2.28), it follows from Theorem 2.26 that the solution $x = x(t)$ of the initial value problem (2.26) with $t_0 = 0$ satisfies

$$\|x(t)\| \le ce^{-dt}\|x_0\| + \int_0^t ce^{-d(t-s)} K \, ds$$

for $t \ge 0$, where $K = \sup_{s>0} \|b(s)\|$. Thus,

$$\|x(t)\| \le ce^{-dt}\|x_0\| + \frac{c}{d}\left(1 - e^{-dt}\right) K$$

for $t \ge 0$. In particular, the solution is bounded in $\mathbb{R}^+$.

## 2.4. Equations with periodic coefficients

In this section we consider the particular case of the linear equations with periodic coefficients. More precisely, we consider equations in $\mathbb{R}^n$ of the form

$$x' = A(t)x, \tag{2.30}$$

where the matrices $A(t) \in M_n$ vary continuously with $t \in \mathbb{R}$, and we assume that there exists $T > 0$ such that

$$A(t + T) = A(t) \tag{2.31}$$

for every $t \in \mathbb{R}$.

**Definition 2.29.** Given $T > 0$, a function $F \colon \mathbb{R} \to \mathbb{R}^m$ is said to be $T$-*periodic* if

$$F(t + T) = F(t) \quad \text{for every} \quad t \in \mathbb{R}.$$

The function $F$ is said to be *periodic* if it is $T$-periodic for some $T > 0$.

According to Definition 2.29, the constant functions are $T$-periodic for every $T > 0$.

**Example 2.30.** Consider the equation

$$x' = a(t)x,$$

where $a \colon \mathbb{R} \to \mathbb{R}$ is a periodic continuous function. The solutions are given by

$$x(t) = \exp\left(\int_{t_0}^t a(s) \, ds\right) x(t_0), \quad t \in \mathbb{R}.$$

For example, when $a(t) = 1$ we have $x(t) = e^{t-t_0}x(t_0)$, and the only periodic solution is the zero function. On the other hand, when $a(t) = \cos t$ we have

$$x(t) = e^{\sin t - \sin t_0} x(t_0),$$

and for $x(t_0) \ne 0$ the solution is a nonconstant $2\pi$-periodic function.

Now we describe the fundamental solutions of the linear equations with periodic coefficients. We continue to denote by $M_n$ the set of $n \times n$ matrices.

**Theorem 2.31** (Floquet). *If $A \colon \mathbb{R} \to M_n$ is a $T$-periodic continuous function, then any fundamental solution of equation (2.30) is of the form*

$$X(t) = P(t)e^{Bt}, \tag{2.32}$$

*where $B$ and $P(t)$ are $n \times n$ matrices for each $t \in \mathbb{R}$, with*

$$P(t+T) = P(t), \quad t \in \mathbb{R}. \tag{2.33}$$

**Proof.** It follows from (2.4) that if $X(t)$ is a fundamental solution of equation (2.30), then

$$X'(t+T) = A(t+T)X(t+T) = A(t)X(t+T) \tag{2.34}$$

for $t \in \mathbb{R}$. Hence, $Y(t) = X(t+T)$ is also a fundamental solution of equation (2.30), and by Proposition 2.9 there exists an invertible $n \times n$ matrix $C$ such that

$$X(t+T) = X(t)C \tag{2.35}$$

for $t \in \mathbb{R}$. On the other hand, since $C$ is invertible, there exists an $n \times n$ matrix $B$ such that $e^{BT} = C$. It can be obtained as follows. Let $S$ be an invertible $n \times n$ matrix such that $S^{-1}CS$ has the Jordan canonical form in (2.12). For each matrix $R_j = \lambda_j \mathrm{Id} + N_j$, with $\lambda_j \neq 0$ (because $C$ is invertible) and $N_j$ an $n_j \times n_j$ matrix as in (2.15), we define

$$\log R_j = \log \left[ \lambda_j \left( \mathrm{Id} + \frac{N_j}{\lambda_j} \right) \right]$$

$$= (\log \lambda_j)\mathrm{Id} + \sum_{m=1}^{n_j-1} \frac{(-1)^{m+1}N_j^m}{m\lambda_j^m},$$

where $\log \lambda_j$ is obtained from branch of the complex logarithm. One can verify that $e^{\log R_j} = R_j$ for each $j$ (see for example [**16**]). This implies that the matrix

$$B = \frac{1}{T}S \begin{pmatrix} \log R_1 & & 0 \\ & \ddots & \\ 0 & & \log R_k \end{pmatrix} S^{-1} \tag{2.36}$$

satisfies $e^{BT} = C$. Now we define $n \times n$ matrices

$$P(t) = X(t)e^{-Bt}$$

for each $t \in \mathbb{R}$. It follows from (2.35) (see also Exercise 2.3) that

$$\begin{aligned} P(t+T) &= X(t+T)e^{-B(t+T)} \\ &= X(t)e^{BT}e^{-B(t+T)} \\ &= X(t)e^{-Bt} = P(t). \end{aligned}$$

This completes the proof of the theorem.                                    □

We note that the matrix $B$ in (2.32) is never unique. Indeed, if $e^{BT} = C$, then

$$\exp\left[(B + m(2\pi i/T)\mathrm{Id})T\right] = C$$

for every $m \in \mathbb{Z}$. We also observe that the matrix $P(t)$ is invertible for every $t \in \mathbb{R}$, because by Proposition 2.7 all matrices $X(t)$ are invertible.

We continue to consider a $T$-periodic continuous function $A(t)$ and we introduce the following notions.

**Definition 2.32.** Given a fundamental solution $X(t)$ of equation (2.30):

a) an invertible matrix $C \in M_n$ such that $X(t + T) = X(t)C$ for every $t \in \mathbb{R}$ is called a *monodromy matrix* of equation (2.30);

b) the eigenvalues of a monodromy matrix $C$ are called *characteristic multipliers* of equation (2.30);

c) a number $\lambda \in \mathbb{C}$ such that $e^{\lambda T}$ is a characteristic multiplier is called a *characteristic exponent* of equation (2.30).

In fact, both the characteristic multipliers and the characteristic exponents are independent of the monodromy matrix that was used to define them. This is an immediate consequence of the following result.

**Proposition 2.33.** *Let $A\colon \mathbb{R} \to M_n$ be a $T$-periodic continuous function. If $X(t)$ and $Y(t)$ are fundamental solutions of equation (2.30) with monodromy matrices, respectively, $C$ and $D$, then there exists an invertible $n \times n$ matrix $S$ such that*

$$S^{-1}CS = D. \tag{2.37}$$

**Proof.** The monodromy matrices $C$ and $D$ satisfy

$$X(t + T) = X(t)C \quad \text{and} \quad Y(t + T) = Y(t)D \tag{2.38}$$

for every $t \in \mathbb{R}$. On the other hand, it follows from (2.34) that $X(t+T)$ and $Y(t + T)$ are also fundamental solutions. Thus, by Proposition 2.9, there exists an invertible $n \times n$ matrix $S$ such that

$$Y(t + T) = X(t + T)S$$

for every $t \in \mathbb{R}$. Therefore,

$$Y(t + T) = X(t)CS = Y(t)S^{-1}CS. \tag{2.39}$$

Comparing (2.39) with the second identity in (2.38) yields (2.37).          □

Now we give some examples of linear equations with periodic coefficients, starting with the particular case of constant coefficients.

**Example 2.34.** When there exists a matrix $A \in M_n$ such that $A(t) = A$ for all $t \in \mathbb{R}$, one can take any constant $T > 0$ in (2.31). It follows from Floquet's theorem (Theorem 2.31) that there exist matrices $B$ and $P(t)$ such that

$$e^{At} = P(t)e^{Bt} \tag{2.40}$$

for every $t \in \mathbb{R}$. Since $At$ and $AT$ commute, we have

$$e^{A(t+T)} = e^{At}e^{AT}$$

(see Exercise 2.3), and thus $C = e^{AT}$ is a monodromy matrix of equation (2.30). Now let $S$ be an invertible $n \times n$ matrix such that $S^{-1}AS$ has the Jordan canonical form in (2.12). Proceeding as in (2.36), one can take

$$\begin{aligned}
B &= \frac{1}{T} S \log(S^{-1} e^{AT} S) S^{-1} \\
&= \frac{1}{T} S \log e^{(S^{-1}AS)T} S^{-1} \\
&= \frac{1}{T} S \begin{pmatrix} \log e^{R_1 T} & & 0 \\ & \ddots & \\ 0 & & \log e^{R_k T} \end{pmatrix} S^{-1} \\
&= \frac{1}{T} S \begin{pmatrix} R_1 T & & 0 \\ & \ddots & \\ 0 & & R_k T \end{pmatrix} S^{-1} \\
&= S \begin{pmatrix} R_1 & & 0 \\ & \ddots & \\ 0 & & R_k \end{pmatrix} S^{-1} = A,
\end{aligned}$$

and it follows from (2.40) that $P(t) = \mathrm{Id}$ for every $t \in \mathbb{R}$.

For example, taking $T = 1$, a monodromy matrix of the equation $x' = Ax$ is $C = e^A$. Hence, the characteristic multipliers are the numbers $e^{\lambda_i} \in \mathbb{C}$, where $\lambda_1, \ldots, \lambda_n$ are the eigenvalues of $A$, and the characteristic exponents are $\lambda_1, \ldots, \lambda_n$, up to an integer multiple of $2\pi i$.

**Example 2.35.** Consider the linear equation with $2\pi$-periodic coefficients

$$\begin{pmatrix} x \\ y \end{pmatrix}' = \begin{pmatrix} 1 & 2 + \sin t \\ 0 & -\cos t/(2 + \sin t) \end{pmatrix} \begin{pmatrix} x \\ y \end{pmatrix}.$$

One can easily verify that the solutions are given by

$$\begin{pmatrix} x(t) \\ y(t) \end{pmatrix} = \begin{pmatrix} -c + de^t \\ c/(2 + \sin t) \end{pmatrix} = \begin{pmatrix} -1 & e^t \\ 1/(2 + \sin t) & 0 \end{pmatrix} \begin{pmatrix} c \\ d \end{pmatrix},$$

with $c, d \in \mathbb{R}$. Hence, a fundamental solution is

$$X(t) = \begin{pmatrix} -1 & e^t \\ 1/(2 + \sin t) & 0 \end{pmatrix},$$

which yields the monodromy matrix

$$C = X(0)^{-1}X(2\pi)$$

$$= \begin{pmatrix} 0 & 2 \\ 1 & 2 \end{pmatrix} \begin{pmatrix} -1 & e^{2\pi} \\ 1/2 & 0 \end{pmatrix} = \begin{pmatrix} 1 & 0 \\ 0 & e^{2\pi} \end{pmatrix}.$$

Therefore, the characteristic multipliers are $1$ and $e^{2\pi}$.

The following result shows that the characteristic exponents determine the asymptotic behavior of the solutions of a linear equation with periodic coefficients.

**Proposition 2.36.** *Let $A \colon \mathbb{R} \to M_n$ be a $T$-periodic continuous function. Then $\lambda$ is a characteristic exponent of equation (2.30) if and only if $e^{\lambda t}p(t)$ is a solution of equation (2.30) for some nonvanishing $T$-periodic function $p$.*

**Proof.** We first assume that $e^{\lambda t}p(t)$ is a solution of equation (2.30) for some nonvanishing $T$-periodic function $p$. It follows from Floquet's theorem (Theorem 2.31) that there exists $x_0 \in \mathbb{R}^n$ such that

$$e^{\lambda t}p(t) = P(t)e^{Bt}x_0$$

for $t \in \mathbb{R}$. Moreover, it follows from (2.33) that

$$e^{\lambda(t+T)}p(t) = P(t)e^{B(t+T)}x_0,$$

that is,

$$e^{\lambda T}P(t)e^{Bt}x_0 = P(t)e^{Bt}e^{BT}x_0.$$

Therefore,

$$P(t)e^{Bt}\big(e^{BT} - e^{\lambda T}\mathrm{Id}\big)x_0 = 0.$$

Since $P(t)e^{Bt}$ is invertible for each $t \in \mathbb{R}$, the number $e^{\lambda T}$ is an eigenvalue of the monodromy matrix $e^{BT}$, and $\lambda$ is a characteristic exponent.

Now we assume that $\lambda$ is a characteristic exponent. Then $e^{\lambda T}$ is an eigenvalue of $e^{BT}$. It follows immediately from Propositions 2.17 and 2.18 that if $\mu$ is an eigenvalue of $B$, then $e^{\mu T}$ is an eigenvalue of $e^{BT}$, with the same multiplicity. This implies that $\lambda$ is an eigenvalue of $B$, up to an integer multiple of $2\pi i$. Hence, there exists $x_0 \in \mathbb{R}^n \setminus \{0\}$ such that $Bx_0 = \lambda x_0$, and

$$e^{Bt}x_0 = e^{\lambda t}x_0$$

for $t \in \mathbb{R}$. Multiplying by $P(t)$, we obtain the solution

$$P(t)e^{Bt}x_0 = e^{\lambda t}P(t)x_0 = e^{\lambda t}p(t),$$

where $p(t) = P(t)x_0$. Clearly, the function $p$ does not vanish, and it follows from (2.33) that

$$p(t+T) = P(t+T)x_0 = P(t)x_0 = p(t).$$

This completes the proof of the proposition. $\qquad\qquad\square$

The following example shows that in general the eigenvalues of the matrices $A(t)$ do not determine the asymptotic behavior of the solutions.

**Example 2.37** (Markus–Yamabe). Consider the equation $x' = A(t)x$ in $\mathbb{R}^2$ with

$$A(t) = \frac{1}{2}\begin{pmatrix} -2 + 3\cos^2 t & 2 - 3\sin t \cos t \\ -2 - 3\sin t \cos t & -2 + 3\sin^2 t \end{pmatrix}.$$

One can easily verify that the eigenvalues of the matrix $A(t)$ are $(-1 \pm i\sqrt{7})/4$ for every $t \in \mathbb{R}$. In particular, they have negative real part. On the other hand, a fundamental solution is

$$X(t) = \begin{pmatrix} e^{t/2}\cos t & e^{-t}\sin t \\ -e^{t/2}\sin t & e^{-t}\cos t \end{pmatrix}, \tag{2.41}$$

and so there exist solutions growing exponentially. Incidentally, by Proposition 2.36, it follows from (2.41) that the characteristic exponents are $1/2$ and $-1$.

Proposition 2.36 also allows one to establish a criterion for the existence of $T$-periodic solutions.

**Proposition 2.38.** *Let $A\colon \mathbb{R} \to M_n$ be a $T$-periodic continuous function. If $1$ is a characteristic multiplier of equation (2.30), then there exists a nonvanishing $T$-periodic solution.*

**Proof.** Let $\lambda \in \mathbb{C}$ be a characteristic exponent such that $e^{\lambda T} = 1$. By Proposition 2.36, there exists a nonvanishing $T$-periodic function $p$ such that $x(t) = e^{\lambda t}p(t)$ is a solution of equation (2.30). We have

$$\begin{aligned} x(t + T) &= e^{\lambda(t+T)}p(t + T) \\ &= e^{\lambda T}e^{\lambda t}p(t) = x(t) \end{aligned}$$

for every $t \in \mathbb{R}$, and the function $x$ is $T$-periodic. $\qquad\square$

Now we describe how one can obtain some information about the characteristic multipliers and the characteristic exponents without solving explicitly the equation.

**Proposition 2.39.** *Let $A\colon \mathbb{R} \to M_n$ be a $T$-periodic continuous function. If the characteristic multipliers of equation (2.30) are $\rho_j = e^{\lambda_j T}$, for $j = 1, \ldots, n$, then*

$$\prod_{j=1}^{n} \rho_j = \exp\int_0^T \operatorname{tr} A(s)\, ds \tag{2.42}$$

*and*

$$\sum_{j=1}^{n} \lambda_j = \frac{1}{T}\int_0^T \operatorname{tr} A(s)\, ds \quad (\operatorname{mod} 2\pi i/T). \tag{2.43}$$

**Proof.** Let $X(t)$ be a fundamental solution of equation (2.30) with monodromy matrix $C$. It follows from (2.35) and Liouville's formula (Theorem 2.10) that

$$\det C = \frac{\det X(t+T)}{\det X(t)}$$

$$= \exp \int_t^{t+T} \operatorname{tr} A(s)\, ds = \exp \int_0^T \operatorname{tr} A(s)\, ds, \tag{2.44}$$

where the last identity is a consequence of the periodicity of $A$. Since the characteristic multipliers are the eigenvalues of $C$, identity (2.42) follows immediately from (2.44). For identity (2.43), we first note that

$$\prod_{j=1}^n \rho_j = \prod_{j=1}^n e^{\lambda_j T} = \exp\left(T \sum_{j=1}^n \lambda_j\right).$$

Thus, it follows from (2.42) that

$$\exp\left(T \sum_{j=1}^n \lambda_j\right) = \exp \int_0^T \operatorname{tr} A(s)\, ds.$$

This yields the desired result. $\qquad\qquad\qquad\qquad\qquad\qquad\qquad\square$

**Example 2.40.** Let $A(t)$ be a $T$-periodic continuous function. We discuss the existence of unbounded solutions in terms of the sign of

$$s = \operatorname{Re} \sum_{j=1}^n \lambda_j,$$

where the numbers $\lambda_j$ are the characteristic exponents. If $s > 0$, then there exists $j$ such that $\operatorname{Re} \lambda_j > 0$, and it follows from Proposition 2.36 that there is an unbounded solution in $\mathbb{R}^+$. On the other hand, if $s < 0$, then there exists $j$ such that $\operatorname{Re} \lambda_j < 0$, and it follows from Proposition 2.36 that there is an unbounded solution in $\mathbb{R}^-$.

**Example 2.41.** Consider the linear equation

$$\begin{cases} x' = x + y, \\ y' = x + (\cos^2 t)y. \end{cases} \tag{2.45}$$

We note that the matrix

$$A(t) = \begin{pmatrix} 1 & 1 \\ 1 & \cos^2 t \end{pmatrix}$$

is $\pi$-periodic. Since

$$\int_0^\pi \operatorname{tr} A(s)\, ds = \int_0^\pi (1 + \cos^2 s)\, ds > 0,$$

it follows from (2.43) that

$$\operatorname{Re}(\lambda_1 + \lambda_2) = \frac{1}{\pi} \int_0^\pi (1 + \cos^2 s)\, ds > 0,$$

where $\lambda_1$ and $\lambda_2$ are the characteristic exponents. Hence, by Example 2.40, there exists a solution of equation (2.45) that is unbounded in $\mathbb{R}^+$.

**Example 2.42.** Consider the equation

$$x'' + p(t)x = 0,$$

where $p \colon \mathbb{R} \to \mathbb{R}$ is a $T$-periodic continuous function. Letting $x' = y$, the equation can be written in the form

$$\begin{pmatrix} x \\ y \end{pmatrix}' = \begin{pmatrix} 0 & 1 \\ -p(t) & 0 \end{pmatrix} \begin{pmatrix} x \\ y \end{pmatrix}. \tag{2.46}$$

We note that the matrix

$$A(t) = \begin{pmatrix} 0 & 1 \\ -p(t) & 0 \end{pmatrix}$$

is $T$-periodic. Since $\operatorname{tr} A(t) = 0$ for every $t \in \mathbb{R}$, it follows from Proposition 2.39 that

$$\rho_1 \rho_2 = \exp \int_0^T \operatorname{tr} A(s)\, ds = 1,$$

where $\rho_1$ and $\rho_2$ are the characteristic multipliers. Taking $\lambda \in \mathbb{C}$ such that $e^{\lambda T} = \rho_1$ and $e^{-\lambda T} = \rho_2$, it follows from Proposition 2.36 that there exist solutions $e^{\lambda t} p_1(t)$ and $e^{-\lambda t} p_2(t)$ of equation (2.46), where $p_1$ and $p_2$ are nonvanishing $T$-periodic functions.

## 2.5. Conjugacies between linear equations

In this section we describe how to compare the solutions of two linear differential equations with constant coefficients. To that effect, we first introduce the notion of conjugacy.

**2.5.1. Notion of conjugacy.** We begin with an example that illustrates the problems in which we are interested.

**Example 2.43.** Consider the scalar equations

$$x' = x \quad \text{and} \quad y' = 2y. \tag{2.47}$$

The solutions are given respectively by

$$x(t) = e^t x(0) \quad \text{and} \quad y' = e^{2t} y(0),$$

and have maximal interval $\mathbb{R}$. The phase portrait of both equations in (2.47) is the one shown in Figure 2.15.

0

**Figure 2.15.** Phase portrait of the equations in (2.47).

This means that from the *qualitative* point of view the two equations in (2.47) cannot be distinguished. However, the speed along the solutions is not the same in both equations. Thus, it is also of interest to compare the solutions from the *quantitative* point of view. More precisely, we want to find a bijective transformation $h \colon \mathbb{R} \to \mathbb{R}$ such that

$$h(e^t x) = e^{2t} h(x) \tag{2.48}$$

for $t, x \in \mathbb{R}$. Identity (2.48) can be described by a commutative diagram. Namely, for each $t \in \mathbb{R}$ we define transformations $\varphi_t, \psi_t \colon \mathbb{R} \to \mathbb{R}$ by

$$\varphi_t(x) = e^t x \quad \text{and} \quad \psi_t(x) = e^{2t} x.$$

One can easily verify that the families of transformations $\varphi_t$ and $\psi_t$ are flows (see Definition 1.11). Identity (2.48) can now be written in the form

$$h \circ \varphi_t = \psi_t \circ h,$$

which corresponds to the commutative diagram

$$
\begin{array}{ccc}
\mathbb{R} & \xrightarrow{\ \varphi_t\ } & \mathbb{R} \\
\ \downarrow{\scriptstyle h} & & \ \downarrow{\scriptstyle h} \\
\mathbb{R} & \xrightarrow[\ \psi_t\ ]{} & \mathbb{R}
\end{array} \ .
$$

When identity (2.48) holds, the solution $x(t) = e^t x(0)$ of the first equation in (2.47) is transformed bijectively onto the solution $y(t) = e^{2t} h(x(0))$ of the second equation. In particular, we must have $h(0) = 0$, since the only critical point of the first equation must be transformed onto the only critical point of the second equation (or $h$ would not transform, bijectively, solutions onto solutions).

Assuming that $h$ is differentiable, one can take derivatives with respect to $t$ in (2.48) to obtain

$$h'(e^t x)e^t x = 2e^{2t} h(x).$$

Taking $t = 0$ yields the identity

$$\frac{h'(x)}{h(x)} = \frac{2}{x} \quad \text{for} \quad x \in \mathbb{R} \setminus \{0\}$$

(we note that $h(x) \neq 0$ for $x \neq 0$, because $h(0) = 0$ and $h$ is bijective). Integrating on both sides, we obtain

$$\log|h(x)| = \log(x^2) + c$$

for some constant $c \in \mathbb{R}$. This is the same as

$$h(x) = ax^2$$

for some constant $a \in \mathbb{R} \setminus \{0\}$, with $x > 0$ or $x < 0$. In order that $h$ is bijective, we take

$$h(x) = \begin{cases} ax^2 & \text{if } x > 0, \\ 0 & \text{if } x = 0, \\ bx^2 & \text{if } x < 0, \end{cases} \tag{2.49}$$

where $a, b \in \mathbb{R}$ are constants such that $ab < 0$. One can easily verify that $h$ is differentiable in $\mathbb{R}$, with $h'(0) = 0$. Incidentally, the inverse of $h$ is not differentiable at the origin.

Now we consider arbitrary linear equations. More precisely, in an analogous manner to that in Example 2.43, we consider the linear equations

$$x' = Ax \quad \text{and} \quad y' = By \tag{2.50}$$

in $\mathbb{R}^n$, where $A$ and $B$ are $n \times n$ matrices. The solutions are, respectively,

$$x(t) = e^{At}x(0) \quad \text{and} \quad y(t) = e^{Bt}y(0),$$

and have maximal interval $\mathbb{R}$.

**Definition 2.44.** The solutions of the equations in (2.50) are said to be:

a) *topologically conjugate* if there is a homeomorphism $h \colon \mathbb{R}^n \to \mathbb{R}^n$ (that is, a bijective continuous transformation with continuous inverse) such that

$$h(e^{At}x) = e^{Bt}h(x) \tag{2.51}$$

for every $t \in \mathbb{R}$ and $x \in \mathbb{R}^n$;

b) *differentially conjugate* if there is a diffeomorphism $h$ (that is, a bijective differentiable transformation with differentiable inverse) such that identity (2.51) holds for every $t \in \mathbb{R}$ and $x \in \mathbb{R}^n$;

c) *linearly conjugate* if there is an invertible linear transformation $h$ such that identity (2.51) holds for every $t \in \mathbb{R}$ and $x \in \mathbb{R}^n$.

We then say, respectively, that $h$ is a *topological*, *differentiable* and *linear conjugacy* between the solutions of the equations in (2.50).

In an analogous manner to that in Example 2.43, for each $t \in \mathbb{R}$ we define transformations $\varphi_t, \psi_t \colon \mathbb{R}^n \to \mathbb{R}^n$ by

$$\varphi_t(x) = e^{At}x \quad \text{and} \quad \psi_t(x) = e^{Bt}x.$$

Then identity (2.51) can be written in the form

$$h \circ \varphi_t = \psi_t \circ h,$$

which corresponds to the commutative diagram

$$
\begin{array}{ccc}
\mathbb{R}^n & \xrightarrow{\;\varphi_t\;} & \mathbb{R}^n \\
{\scriptstyle h}\downarrow & & \downarrow{\scriptstyle h} \\
\mathbb{R}^n & \xrightarrow[\;\psi_t\;]{} & \mathbb{R}^n
\end{array}
\quad.
$$

Clearly, if the solutions of the equations in (2.50) are linearly conjugate, then they are differentially conjugate (since any invertible linear transformation is a diffeomorphism), and if the solutions are differentially conjugate, then they are topologically conjugate (since any diffeomorphism is a homeomorphism).

**2.5.2. Linear conjugacies.** In this section we describe some relations between the various notions of conjugacy. We first show that the notions of linear conjugacy and differentiable conjugacy are equivalent for the linear equations in (2.50).

**Proposition 2.45.** *The solutions of the equations in (2.50) are differentially conjugate if and only if they are linearly conjugate.*

**Proof.** As we observed earlier, if the solutions are linearly conjugate, then they are differentially conjugate.

Now let $h\colon \mathbb{R}^n \to \mathbb{R}^n$ be a differentiable conjugacy. Taking derivatives in identity (2.51) with respect to $x$, we obtain

$$d_{e^{At}x}h e^{At} = e^{Bt}d_x h,$$

where $d_x h$ is the Jacobian matrix of $h$ at the point $x$. Taking $x = 0$ yields the identity

$$Ce^{At} = e^{Bt}C, \qquad (2.52)$$

where $C = d_0 h$. We note that $C$ is an invertible $n \times n$ matrix. Indeed, taking derivatives in the identity $h^{-1}(h(x)) = x$, we obtain

$$d_{h(0)}h^{-1}C = \mathrm{Id},$$

which shows that the matrix $C$ is invertible. For the invertible linear transformation $g\colon \mathbb{R}^n \to \mathbb{R}^n$ given by $g(x) = Cx$, it follows from (2.52) that

$$g(e^{At}x) = e^{Bt}g(x)$$

for every $t \in \mathbb{R}$ and $x \in \mathbb{R}^n$. In other words, $g$ is a linear conjugacy. $\qquad\square$

By Proposition 2.45, in what concerns the study of conjugacies for linear equations it is sufficient to consider linear conjugacies and topological conjugacies.

**Example 2.46** (continuation of Example 2.43). We already showed that the bijective differentiable transformations $h\colon \mathbb{R} \to \mathbb{R}$ that are topological conjugacies between the solutions of the equations in (2.47) take the form (2.49) with $ab < 0$. Since none of these functions is linear, there are no linear conjugacies and it follows from Proposition 2.45 that there are also no differentiable conjugacies. Indeed, as we already observed in Example 2.43, the inverse of the function $h$ in (2.49) is not differentiable.

Now we establish a criterion for the existence of a linear conjugacy in terms of the matrices $A$ and $B$ in (2.50).

**Proposition 2.47.** *The solutions of the equations in* (2.50) *are linearly conjugate if and only if there exists an invertible $n \times n$ matrix $C$ such that*

$$A = C^{-1}BC. \tag{2.53}$$

**Proof.** We first assume that the solutions are linearly conjugate, that is, identity (2.51) holds for some invertible linear transformation $h(x) = Cx$, where $C$ is thus an invertible matrix. Then identity (2.52) holds and taking derivatives with respect to $t$, we obtain

$$CAe^{At} = Be^{Bt}C.$$

Finally, taking $t = 0$ yields $CA = BC$, and identity (2.53) holds.

Now we assume that there exists an invertible $n \times n$ matrix $C$ as in (2.53). Then

$$x' = Ax \quad \Leftrightarrow \quad x' = C^{-1}BCx \quad \Leftrightarrow \quad (Cx)' = B(Cx).$$

Solving the first and third equations, we obtain

$$x(t) = e^{At}x(0) \qquad \text{and} \qquad Cx(t) = e^{Bt}Cx(0)$$

for every $t \in \mathbb{R}$ and $x(0) \in \mathbb{R}^n$. Therefore,

$$Ce^{At}x(0) = e^{Bt}Cx(0),$$

and since the vector $x(0)$ is arbitrary we obtain (2.52). Hence, identity (2.51) holds for the invertible linear transformation $h(x) = Cx$. $\qquad\square$

It follows from Proposition 2.47 that the solutions of the equations in (2.50) are linearly conjugate if and only if $A$ and $B$ have the same Jordan canonical form.

**2.5.3. Topological conjugacies.** In this section we consider the notion of topological conjugacy. Since it is not as stringent as the notions of differentiable conjugacy and linear conjugacy, it classifies less equations. In other words, the equivalence classes obtained from the notion of topological conjugacy have more elements. However, for the same reason, it allows us to compare linear equations that otherwise could not be compared. For example, one can obtain topological conjugacies between the solutions of equations whose matrices $A$ and $B$ have different Jordan canonical forms, in contrast to what happens in Proposition 2.47 with the notion of linear conjugacy.

A detailed study of topological conjugacies falls outside the scope of the book, and we consider only a particular class of matrices.

**Definition 2.48.** A square matrix $A$ is said to be *hyperbolic* if all of its eigenvalues have nonzero real part.

We recall that the eigenvalues of a matrix vary continuously with its entries (for a proof based on Rouché's theorem see, for example, [**5**]). Thus, a matrix obtained from a sufficiently small perturbation (of the entries of) a hyperbolic matrix is still hyperbolic.

**Definition 2.49.** For a square matrix $A$, we denote by $m(A)$ the number of eigenvalues of $A$ with positive real part, counted with their multiplicities.

Now we establish a criterion for the existence of topological conjugacies between equations with hyperbolic matrices.

**Theorem 2.50.** *Let $A$ and $B$ be hyperbolic $n \times n$ matrices. If*

$$m(A) = m(B), \qquad (2.54)$$

*then the solutions of the equations in (2.50) are topologically conjugate.*

**Proof.** We first assume that $m(A) = 0$. We then explain how one can obtain the result from this particular case. We construct explicitly topological conjugacies as follows.

*Step 1. Construction of an auxiliary function.* Define a function $q \colon \mathbb{R}^n \to \mathbb{R}$ by

$$q(x) = \int_0^\infty \|e^{At}x\|^2 \, dt,$$

where $\|\cdot\|$ is the norm in $\mathbb{R}^n$ given by

$$\|(x_1, \ldots, x_n)\| = \left( \sum_{i=1}^n x_i^2 \right)^{1/2}.$$

Since $A$ has only eigenvalues with negative real part, by Proposition 2.27 there exist constants $c, \mu > 0$ such that $\|e^{At}\| \leq ce^{-\mu t}$ for every $t \geq 0$. This implies that

$$\int_0^\infty \|e^{At}x\|^2 \, dt \leq \int_0^\infty c^2 e^{-2\mu t}\|x\|^2 \, dt < +\infty,$$

and thus, the function $q$ is well defined. Denoting by $B^*$ the transpose of a matrix $B$, we observe that

$$\begin{aligned} q(x) &= \int_0^\infty (e^{At}x)^* e^{At}x \, dt \\ &= \int_0^\infty x^*(e^{At})^* e^{At}x \, dt = x^* C x, \end{aligned} \tag{2.55}$$

where

$$C = \int_0^\infty (e^{At})^* e^{At} \, dt$$

is an $n \times n$ matrix. This shows that $q$ is a polynomial of degree 2 in $\mathbb{R}^n$, without terms of degree 0 or 1. In particular,

$$q\left(\frac{x}{\|x\|}\right) = \frac{q(x)}{\|x\|^2} \tag{2.56}$$

for $x \neq 0$. Thus, taking

$$\alpha = \min\{q(x) : \|x\| = 1\} \quad \text{and} \quad \beta = \max\{q(x) : \|x\| = 1\},$$

it follows from (2.56) that

$$\alpha\|x\|^2 \leq q(x) \leq \beta\|x\|^2. \tag{2.57}$$

*Step 2. Construction of the conjugacy.* Let $x \neq 0$. We note that

$$\|e^{As}x\| \to 0 \quad \text{when} \quad s \to +\infty$$

and

$$\|e^{As}x\| \to +\infty \quad \text{when} \quad s \to -\infty. \tag{2.58}$$

The first property follows immediately from Proposition 2.27. For the second property, we note that the matrix $e^{As}$ is nonsingular (see Exercise 2.3), and hence $e^{As}x \neq 0$ for $x \neq 0$. Now we consider the root spaces of $A$, that is,

$$F_\lambda = \{x \in \mathbb{C}^n : (A - \lambda\operatorname{Id})^k x = 0 \text{ for some } k \in \mathbb{N}\}$$

for each eigenvalue $\lambda$ of $A$. We recall that $\mathbb{C}^n = \bigoplus_\lambda F_\lambda$, where the direct sum is taken over all eigenvalues of $A$, counted without their multiplicities. Hence,

$$\mathbb{R}^n = \bigoplus_\lambda (\mathbb{R}^n \cap F_\lambda).$$

In particular, given $x \in \mathbb{R}^n \setminus \{0\}$, there exists at least one eigenvalue $\lambda$ of $A$ such that the component of $x$ in $\mathbb{R}^n \cap F_\lambda$ is nonzero. Thus, by (2.16), in order to establish property (2.58) it is sufficient to show that all nonzero entries of the matrix

$$e^{Bt} = e^{\lambda t}\left(\mathrm{Id} + tN + \cdots + \frac{1}{(m-1)!}t^{m-1}N^{m-1}\right),$$

where $B = \lambda\mathrm{Id} + N$ is an $m \times m$ Jordan block, have the same property. To that effect, we note that if $\lambda$ has negative real part and $p(t)$ is a nonzero polynomial, then

$$|e^{\lambda t}p(t)| \to +\infty \quad \text{when} \quad t \to -\infty.$$

We proceed with the proof of the theorem. It follows from (2.57) that the image of the function $s \mapsto q(e^{As}x)$ is $\mathbb{R}^+$. On the other hand, we have

$$q(e^{As}x) = \int_0^\infty \|e^{A(t+s)}x\|^2\,dt = \int_s^\infty \|e^{At}x\|^2\,dt, \tag{2.59}$$

and thus, the function $s \mapsto q(e^{As}x)$ is strictly decreasing. In particular, there exists a unique $t_x \in \mathbb{R}$ such that $q(e^{At_x}x) = 1$. Now we define a transformation $h\colon \mathbb{R}^n \to \mathbb{R}^n$ by

$$h(x) = \begin{cases} e^{-Bt_x}e^{At_x}x/\bar{q}(e^{At_x}x)^{1/2} & \text{if } x \neq 0, \\ 0 & \text{if } x = 0, \end{cases} \tag{2.60}$$

where

$$\bar{q}(x) = \int_0^\infty \|e^{Bt}x\|^2\,dt. \tag{2.61}$$

Since

$$q(e^{A(t_x-t)}e^{At}x) = q(e^{At_x}x) = 1,$$

we have $t_{e^{At}x} = t_x - t$. Therefore,

$$\begin{aligned} h(e^{At}x) &= \frac{e^{-B(t_x-t)}e^{A(t_x-t)}e^{At}x}{\bar{q}(e^{A(t_x-t)}e^{At}x)^{1/2}} \\ &= \frac{e^{Bt}e^{-Bt_x}e^{At_x}x}{\bar{q}(e^{At_x}x)^{1/2}} = e^{Bt}h(x), \end{aligned}$$

which establishes identity (2.51). In order to show that $h$ is a topological conjugacy, it remains to verify that $h$ is a homeomorphism.

*Step 3. Continuity of $h$.* It follows from (2.59) and the Implicit function theorem that the function $x \mapsto t_x$ is differentiable (outside the origin). On the other hand, proceeding as in (2.55), one can show that $\bar{q}$ is a polynomial of degree 2 in $\mathbb{R}^n$, without terms of degree 0 or 1. Thus, it follows from (2.60) that the transformation $h$ is differentiable outside the origin. Now we show that $h$ is continuous at $x = 0$. Proceeding as in (2.56) and taking

$$\bar{\alpha} = \min\{\bar{q}(x) : \|x\| = 1\} \quad \text{and} \quad \bar{\beta} = \max\{\bar{q}(x) : \|x\| = 1\},$$

we obtain

$$\bar{\alpha}\|x\|^2 \le \bar{q}(x) \le \bar{\beta}\|x\|^2. \tag{2.62}$$

By (2.62), in order to show that $h$ is continuous at the origin it is sufficient to prove that $\bar{q}(h(x)) \to 0$ when $x \to 0$. By (2.60), for $x \ne 0$ we have

$$\bar{q}(h(x)) = \frac{\bar{q}(e^{-Bt_x}e^{At_x}x)}{\bar{q}(e^{At_x}x)}.$$

On the other hand, it follows from (2.61) that

$$\bar{q}(e^{-Bt_x}e^{At_x}x) = \int_0^\infty \|e^{-Bt_x}e^{Bt}e^{At_x}x\|^2 \, dt \le \|e^{-Bt_x}\|^2 \bar{q}(e^{At_x}x),$$

and hence,

$$\bar{q}(h(x)) \le \|e^{-Bt_x}\|^2. \tag{2.63}$$

Now we observe that, by (2.59),

$$q(e^{At_x}x) = \int_{t_x}^\infty \|e^{At}x\|^2 \, dt = 1.$$

Thus, $t_x \to -\infty$ when $x \to 0$. Since $B$ has only eigenvalues with negative real part, it follows from (2.63) that $\bar{q}(h(x)) \to 0$ when $x \to 0$. This shows that $h$ is continuous in $\mathbb{R}^n$.

*Step 4. Construction of the inverse of $h$.* We define a transformation $g\colon \mathbb{R}^n \to \mathbb{R}^n$ by

$$g(x) = \begin{cases} e^{-As_x}e^{Bs_x}x/q(e^{Bs_x}x)^{1/2} & \text{if } x \ne 0, \\ 0 & \text{if } x = 0, \end{cases}$$

where $s_x$ is the unique real number such that

$$\bar{q}(e^{Bs_x}x) = \int_{s_x}^\infty \|e^{Bt}x\|^2 \, dt = 1$$

(it can be shown in a similar manner to that for $t_x$ that the number $s_x$ exists and is unique). We have

$$\bar{q}(e^{Bt_x}h(x)) = \int_0^\infty \frac{\|e^{Bt}e^{Bt_x}e^{-Bt_x}e^{At_x}x\|^2}{\bar{q}(e^{At_x}x)} \, dt = 1,$$

and the uniqueness of $s_{h(x)}$ implies that $s_{h(x)} = t_x$. Thus, for $x \ne 0$ we have

$$g(h(x)) = \frac{e^{-At_x}e^{Bt_x}e^{-Bt_x}e^{At_x}x}{q(e^{Bt_x}h(x))^{1/2}\bar{q}(e^{At_x}x)^{1/2}}$$

$$= \frac{x}{q(e^{Bt_x}h(x))^{1/2}\bar{q}(e^{At_x}x)^{1/2}}.$$

Since

$$q(e^{Bt_x} h(x)) = q\left(\frac{e^{At_x} x}{\bar{q}(e^{At_x} x)^{1/2}}\right)$$

$$= \frac{q(e^{At_x} x)}{\bar{q}(e^{At_x} x)} = \frac{1}{\bar{q}(e^{At_x} x)},$$

we obtain $g(h(x)) = x$ for $x \neq 0$. We also have $g(h(0)) = g(0) = 0$, which shows that $g$ is the inverse of $h$. In order to show that $g$ is continuous one can proceed as for $h$, simply interchanging the roles of $A$ and $B$. Summing up, $h$ is a homeomorphism.

*Step 5. Reduction to the case $m(A) = 0$.* Finally, we describe how to reduce the general case to the case of matrices with $m(A) = 0$. Making appropriate changes of coordinates, one can always assume that the matrices $A$ and $B$ have, respectively, the forms

$$\begin{pmatrix} A_+ & 0 \\ 0 & A_- \end{pmatrix} \quad \text{and} \quad \begin{pmatrix} B_+ & 0 \\ 0 & B_- \end{pmatrix}, \quad (2.64)$$

where the indices $+$ correspond to the eigenvalues with positive real part and the indices $-$ correspond to the eigenvalues with negative real part. It follows from (2.54) that the matrices $A_+$ and $B_+$ have the same dimension, say $n_+$, and that the matrices $A_-$ and $B_-$ also have the same dimension, say $n_-$. We write $x = (x_+, x_-)$, with $x_+ \in \mathbb{R}^{n_+}$ and $x_- \in \mathbb{R}^{n_-}$. By (2.64), the equations in (2.50) can be written, respectively, in the forms

$$\begin{cases} x'_+ = A_+ x_+, \\ x'_- = A_- x_- \end{cases} \quad \text{and} \quad \begin{cases} x'_+ = B_+ x_+, \\ x'_- = B_- x_-. \end{cases} \quad (2.65)$$

By the result for $m(A) = 0$ and the corresponding result for eigenvalues with positive real part, there exist topological conjugacies $h_+$ and $h_-$, respectively, between the solutions of the equations

$$x'_+ = A_+ x_+ \quad \text{and} \quad x'_+ = B_+ x_+,$$

and between the solutions of the equations

$$x'_- = A_- x_- \quad \text{and} \quad x'_- = B_- x_-.$$

One can easily verify that the transformation $h \colon \mathbb{R}^n \to \mathbb{R}^n$ defined by

$$h(x_+, x_-) = (h_+(x_+), h_-(x_-)) \quad (2.66)$$

is a topological conjugacy between the solutions of the equations in (2.65). $\square$

The following is an application of Theorem 2.50.

**Example 2.51.** Consider the matrices

$$A_1 = \begin{pmatrix} -1 & 0 \\ 0 & -2 \end{pmatrix}, \ A_2 = \begin{pmatrix} -1 & 0 \\ 0 & -1 \end{pmatrix}, \ A_3 = \begin{pmatrix} -1 & 1 \\ -1 & -1 \end{pmatrix}, \ A_4 = \begin{pmatrix} -1 & 0 \\ 0 & 2 \end{pmatrix}.$$

The corresponding linear equations $x' = A_i x$ in $\mathbb{R}^2$ have the phase portraits shown in Figure 2.16. One can easily verify that the four matrices $A_i$ are hyperbolic, with $m(A_i)$, respectively, equal to $0, 0, 0$ and $1$ for $i = 1, 2, 3, 4$. Hence, it follows from Theorem 2.50 that the solutions of the first three equations are topologically conjugate and that they are not topologically conjugate to the solutions of the last equation. On the other hand, since the four matrices have different Jordan canonical forms, it follows from Propositions 2.45 and 2.47 that the solutions of these equations are neither differentially conjugate nor linearly conjugate.

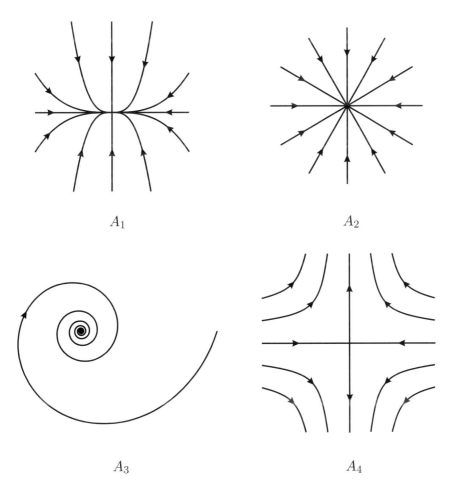

**Figure 2.16.** Phase portraits of the equations in Example 2.51.

The proof of Theorem 2.50 can be used to construct topological conjugacies explicitly. We describe briefly the construction. For each $x \neq 0$, let $t_x \in \mathbb{R}$ be the unique real number such that

$$\int_{t_x}^{\infty} \|e^{At}x\|^2 \, dt = 1 \qquad (2.67)$$

(it is shown in the proof of the theorem that $t_x$ is well defined). Then a topological conjugacy $h \colon \mathbb{R}^n \to \mathbb{R}^n$ between the solutions of the equations $x' = Ax$ and $y' = By$ is given by (2.60), that is,

$$h(x) = \begin{cases} e^{-Bt_x} e^{At_x} x \big/ \left( \int_0^{\infty} \|e^{Bt} e^{At_x} x\|^2 \, dt \right)^{1/2} & \text{if } x \neq 0, \\ 0 & \text{if } x = 0. \end{cases} \qquad (2.68)$$

The following is a specific example.

**Example 2.52.** Consider the matrices

$$A = \begin{pmatrix} -1 & 0 \\ 0 & -1 \end{pmatrix} \qquad \text{and} \qquad B = \begin{pmatrix} -1 & 1 \\ 0 & -1 \end{pmatrix}.$$

We have

$$e^{At} = \begin{pmatrix} e^{-t} & 0 \\ 0 & e^{-t} \end{pmatrix} \qquad \text{and} \qquad e^{Bt} = \begin{pmatrix} e^{-t} & te^{-t} \\ 0 & e^{-t} \end{pmatrix}.$$

Writing $x = (y, z)$, we obtain

$$\|e^{At}x\|^2 = e^{-2t}(y^2 + z^2),$$

and it follows from (2.67) that

$$\int_{t_x}^{\infty} \|e^{At}x\|^2 \, dt = \frac{1}{2} e^{-2t_x}(y^2 + z^2) = 1.$$

Thus,

$$t_x = \frac{1}{2} \log \frac{y^2 + z^2}{2}$$

and

$$e^{-Bt_x} e^{At_x} x = \begin{pmatrix} 1 & t_x \\ 0 & 1 \end{pmatrix} \begin{pmatrix} y \\ z \end{pmatrix} = \begin{pmatrix} y + \frac{z}{2} \log \frac{y^2+z^2}{2} \\ z \end{pmatrix}.$$

In order to determine $h$, we note that

$$\int_0^\infty \|e^{Bt}e^{At_x}x\|^2\,dt = \int_0^\infty \left\|e^{-t-t_x}\begin{pmatrix} y+tz \\ z \end{pmatrix}\right\|^2\,dt$$

$$= e^{-2t_x}\int_0^\infty e^{-2t}(y^2 + 2tyz + t^2z^2 + z^2)\,dt$$

$$= e^{-2t_x}\left(\frac{1}{2}y^2 + \frac{1}{2}yz + \frac{3}{4}z^2\right)$$

$$= \frac{1}{y^2+z^2}\left(y^2 + yz + \frac{3}{2}z^2\right).$$

Finally, by (2.68), we take $h(0,0) = 0$ and

$$h(y,z) = \sqrt{\frac{y^2+z^2}{y^2+yz+3z^2/2}}\left(y + \frac{z}{2}\log\frac{y^2+z^2}{2}, z\right)$$

for $(y,z) \neq (0,0)$.

The problem of the existence of topological conjugacies is very different in the case of nonhyperbolic matrices, as the following example illustrates.

**Example 2.53.** Consider the equations

$$\begin{cases} x' = -ay, \\ y' = ax \end{cases} \quad \text{and} \quad \begin{cases} x' = -by, \\ y' = by \end{cases}$$

for some constants $a, b > 0$. One can easily verify that they have the same phase portrait; namely, the origin is a critical point and the remaining orbits are circular periodic orbits centered at the origin and traversed in the negative direction. However, when $a \neq b$ the periods of the periodic orbits are different in the two equations. Namely, in polar coordinates the equations take respectively the forms

$$\begin{cases} r' = 0, \\ \theta' = a \end{cases} \quad \text{and} \quad \begin{cases} r' = 0, \\ \theta' = b, \end{cases}$$

and thus, the periods of the periodic orbits are, respectively, $2\pi/a$ and $2\pi/b$. When $a \neq b$, this prevents the existence of a topological conjugacy between the solutions of the two equations, because it would have to transform periodic orbits onto periodic orbits of the same period.

## 2.6. Exercises

**Exercise 2.1.** Compute $e^{At}$ for the matrix

$$A = \begin{pmatrix} 0 & -2 \\ 1 & 0 \end{pmatrix}.$$

**Exercise 2.2.** Find necessary and sufficient conditions in terms of a matrix $A \in M_n$ such that for the equation $x' = Ax$:

    a) all solutions are bounded;

    b) all solutions are bounded for $t > 0$;

    c) all solutions converge to the origin.

**Exercise 2.3.** For a matrix $A \in M_n$, consider the equation $x' = Ax$.

    a) Use (2.10) to show that $e^{At}e^{-At} = \mathrm{Id}$ for each $t \in \mathbb{R}$. Hint: Compute the derivative of the function $t \mapsto e^{At}e^{-At}$.

    b) Show that

$$e^{A(t-s)} = e^{At}e^{-As} \quad \text{for every} \quad t, s \in \mathbb{R}. \tag{2.69}$$

    Hint: Take derivatives with respect to $t$.

    c) Use Theorem 2.25 and identity (2.69) to give an alternative proof of Theorem 2.26.

    d) Show that $\det e^A = e^{\mathrm{tr}\, A}$.

**Exercise 2.4.** Given matrices $A, B \in M_n$, show that if

$$[A, [A, B]] = [B, [A, B]] = 0,$$

where $[A, B] = BA - AB$, then

$$e^{At}e^{Bt} = e^{(A+B)t}e^{[A,B]t^2/2}, \quad t \in \mathbb{R}.$$

Hint: Show that

$$x(t) = e^{-(A+B)t}e^{Bt}e^{At}x_0$$

is a solution of the equation $x' = t[A, B]x$ for each $x_0 \in \mathbb{R}^n$.

**Exercise 2.5.** Given a matrix $A \in M_n$, let

$$\cos A = \frac{e^{iA} + e^{-iA}}{2} \quad \text{and} \quad \sin A = \frac{e^{iA} - e^{-iA}}{2i}.$$

Compute these functions for the matrices

$$B = \begin{pmatrix} 0 & 1 & 0 \\ 0 & 0 & 1 \\ 0 & 0 & 0 \end{pmatrix} \quad \text{and} \quad C = \begin{pmatrix} 0 & 0 & 0 \\ 1 & 0 & 0 \\ 0 & 1 & 0 \end{pmatrix}.$$

**Exercise 2.6.** Consider the equation $x' = a(t)x$ for a continuous function $a: \mathbb{R} \to \mathbb{R}$.

    a) Find all solutions of the equation.

    b) Identify the following statement as true or false: There exists a non-vanishing $T$-periodic solution if and only if $\int_0^T a(s)\, ds = 0$.

c) Identify the following statement as true or false: There exists an unbounded solution in $\mathbb{R}^+$ if and only if $\int_0^t a(s)\, ds \neq 0$ for some $t > 0$.

**Exercise 2.7.** For equation (2.1), show that if $\operatorname{tr} A(t) = 0$ for every $t \in \mathbb{R}$, then given $n$ linearly independent solutions $x_1, \ldots, x_n$, the volume determined by the vectors $x_1(x), \ldots, x_n(t)$ is independent of $t$.

**Exercise 2.8.** Given continuous functions $f, g \colon \mathbb{R} \to \mathbb{R}$, solve the equation

$$\begin{pmatrix} x \\ y \end{pmatrix}' = \begin{pmatrix} f(t) & g(t) \\ g(t) & f(t) \end{pmatrix} \begin{pmatrix} x \\ y \end{pmatrix}.$$

**Exercise 2.9.** Let $a, b \colon \mathbb{R} \to \mathbb{R}$ be $T$-periodic continuous functions. Show that if the equation $x' = a(t)x$ has no $T$-periodic solutions other than the zero function, then the equation $x' = a(t)x + b(t)$ has a unique $T$-periodic solution.

**Exercise 2.10.** Show that the equation

$$\begin{cases} x' = x \cos^4 t - z \sin(2t), \\ y' = x \sin(4t) + y \sin t - 4z, \\ z' = -x \sin(5t) - z \cos t \end{cases}$$

has at least one unbounded solution.

**Exercise 2.11.** Show that there exist functions $f, g \colon \mathbb{R}^2 \to \mathbb{R}$ such that the equation

$$\begin{cases} x' = f(x, y), \\ y' = g(x, y) \end{cases} \tag{2.70}$$

has the phase portrait in Figure 2.17.

**Exercise 2.12.** Construct topological conjugacies between the solutions of the equations:

a) $\begin{cases} x' = 2x, \\ y' = -4y \end{cases}$ and $\begin{cases} x' = 3x, \\ y' = -y; \end{cases}$

b) $\begin{cases} x' = -2x, \\ y' = -2y, \\ z' = 3z \end{cases}$ and $\begin{cases} x' = -2x - y, \\ y' = -2y, \\ z' = 2z. \end{cases}$

**Exercise 2.13.** Let $A \colon \mathbb{R}_0^+ \to M_n$ be a continuous function and let $x(t)$ be a solution of the equation $x' = A(t)x$. Show that:

a) for each $t \geq 0$,

$$\|x(t)\| \leq \|x(0)\| \exp \int_0^t \|A(s)\|\, ds;$$

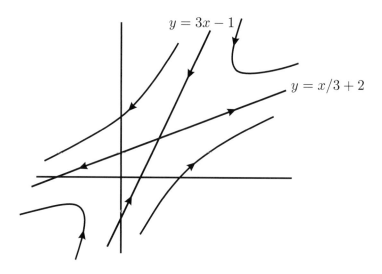

**Figure 2.17.** Phase portrait of equation (2.70).

    b) if $\int_0^\infty \|A(s)\|\, ds < \infty$, then $\|x(t)\|$ converges when $t \to +\infty$.

**Exercise 2.14.** Show that if $x(t)$ and $y(t)$ are, respectively, solutions of the equations

$$x' = A(t)x \quad \text{and} \quad y' = -A(t)^* y$$

in $\mathbb{R}^n$, then $\langle x(t), y(t) \rangle$ is independent of $t$, where $\langle \cdot, \cdot \rangle$ is the usual inner product in $\mathbb{R}^n$.

**Exercise 2.15.** Show that if $X(t)$ is a fundamental solution of the equation $x' = A(t)x$, then $Y(t) = (X(t)^*)^{-1}$ is a fundamental solution of the equation $y' = -A(t)^* y$.

**Exercise 2.16.** Show that if $A(t) = -A(t)^*$ for every $t \in \mathbb{R}$, then $\|x\|^2$ is an integral of the equation $x' = A(t)x$.

**Exercise 2.17.** Verify that the function $h$ in (2.66) is a topological conjugacy.

**Exercise 2.18.** Let $f \colon \mathbb{R} \times \mathbb{R}^n \times \mathbb{R}^m \to \mathbb{R}^n$ be a function of class $C^1$ such that

$$f(t + T, x, \lambda) = f(t, x, \lambda)$$

for every $(t, x, \lambda) \in \mathbb{R} \times \mathbb{R}^n \times \mathbb{R}^m$. Assuming that $x(t)$ is a $T$-periodic solution of the equation $x' = f(t, x, 0)$ and that the only $T$-periodic solution of the linear variational equation

$$y' = \frac{\partial f}{\partial x}(t, x(t), 0)y$$

is the zero function, show that given $\varepsilon > 0$, there exists $\delta > 0$ such that if $\|\lambda\| < \delta$, then there exists a unique $T$-periodic solution $x_\lambda(t)$ of the equation $x' = f(t, x, \lambda)$ satisfying

$$\|x_\lambda(t) - x(t)\| < \varepsilon \quad \text{for} \quad t \in \mathbb{R}.$$

**Solutions.**

**2.1** $e^{At} = \begin{pmatrix} \cos(\sqrt{2}t) & \sqrt{2}\sin(\sqrt{2}t) \\ -\sin(\sqrt{2}t)/\sqrt{2} & \cos(\sqrt{2}t) \end{pmatrix}.$

**2.2** a) $A$ has only eigenvalues with zero real part and diagonal Jordan block.

b) $A$ has no eigenvalues with positive real part and each eigenvalue with zero real part has diagonal Jordan block.

c) $A$ has only eigenvalues with negative real part.

**2.5** $\cos B = \begin{pmatrix} 1 & 0 & -1/2 \\ 0 & 1 & 0 \\ 0 & 0 & 1 \end{pmatrix}$, $\sin B = \begin{pmatrix} 0 & 1 & 0 \\ 0 & 0 & 1 \\ 0 & 0 & 0 \end{pmatrix}$,

$\cos C = \begin{pmatrix} 1 & 0 & 0 \\ 0 & 1 & 0 \\ -1/2 & 0 & 1 \end{pmatrix}$, $\sin C = \begin{pmatrix} 0 & 0 & 0 \\ 1 & 0 & 0 \\ 0 & 1 & 0 \end{pmatrix}.$

**2.6** a) $x(t) = e^{\int_{t_0}^t a(\tau)\,d\tau} x(t_0).$

b) False.

c) False.

**2.8** $x(t) = a \exp \int_0^t (f(s) + g(s))\,ds + b \exp \int_0^t (f(s) - g(s))\,ds,$

$y(t) = a \exp \int_0^t (f(s)+g(s))\,ds - b \exp \int_0^t (f(s)-g(s))\,ds,$ with $a, b \in \mathbb{R}.$

**2.12** a) $h(x, y) = (\operatorname{sgn} x \cdot |x|^{3/2}, \operatorname{sgn} y \cdot |y|^{1/4}).$

b) $h(x, y, z) = (f(x, y), g(z))$, where $g(z) = \operatorname{sgn} z \cdot |z|^{2/3}$ and

$$f(x, y) = \begin{cases} 0 & \text{if } (x, y) = (0, 0), \\ \sqrt{\dfrac{x^2+y^2}{x^2+9y^2/8-4xy/8}}\left(x - \dfrac{y}{4}\log\dfrac{x^2+y^2}{4}, y\right) & \text{if } (x, y) \neq (0, 0). \end{cases}$$

*Part 2*

# Stability and Hyperbolicity

# Chapter 3

# Stability and Lyapunov Functions

In this chapter we introduce the notions of stability and asymptotic stability for a solution of an ordinary differential equation. In particular, for nonautonomous linear equations we characterize the notions of stability and asymptotic stability in terms of the fundamental solutions. We also show that the solutions of a sufficiently small perturbation of an asymptotically stable linear equation remain asymptotically stable. Finally, we give an introduction to the theory of Lyapunov functions, which sometimes allows one to establish in a more or less automatic manner the stability or instability of a given solution. For additional topics we refer the reader to [**20, 24, 25**].

## 3.1. Notions of stability

Given a continuous function $f\colon D \to \mathbb{R}^n$ in an open set $D \subset \mathbb{R} \times \mathbb{R}^n$, consider the equation

$$x' = f(t, x). \tag{3.1}$$

We assume that for each $(t_0, x_0) \in D$ there exists a unique solution $x(t, t_0, x_0)$ of the initial value problem

$$\begin{cases} x' = f(t, x), \\ x(t_0) = x_0. \end{cases}$$

**3.1.1. Stability.** In this section we introduce the notion of stability for a solution of equation (3.1). Essentially, a solution $x(t)$ is stable if all solutions with sufficiently close initial condition remain close to $x(t)$ for all time.

**Definition 3.1.** A solution $x(t, t_0, \bar{x}_0)$ of equation (3.1) defined for all $t > t_0$ is said to be *stable* if given $\varepsilon > 0$, there exists $\delta > 0$ such that if $\|x_0 - \bar{x}_0\| < \delta$, then:

a) the solution $x(t, t_0, x_0)$ is defined for all $t > t_0$;

b) $\|x(t, t_0, x_0) - x(t, t_0, \bar{x}_0)\| < \varepsilon$ for $t > t_0$.

Otherwise, the solution $x(t, t_0, \bar{x}_0)$ is said to be *unstable*.

The following are examples of stability and instability of solutions.

**Example 3.2.** Consider the equation

$$\begin{cases} x' = y, \\ y' = -x - y. \end{cases} \tag{3.2}$$

If $(x, y)$ is a solution, then

$$(x^2 + y^2)' = 2xy + 2y(-x - y) = -2y^2 \le 0,$$

and thus, the conditions in Definition 3.1 are satisfied for the zero solution (with $t_0$ arbitrary and $\bar{x}_0 = 0$). This shows that the critical point $(0,0)$ is a stable solution. Alternatively, note that the matrix of coefficients of the linear equation (3.2) has eigenvalues $(-1 \pm i\sqrt{3})/2$, both with negative real part.

**Example 3.3.** Consider the equation in polar coordinates

$$\begin{cases} r' = 0, \\ \theta' = f(r), \end{cases} \tag{3.3}$$

where $f$ is a positive function of class $C^1$ with $f'(r_0) \ne 0$ for some $r_0 > 0$. The phase portrait is the one shown in Figure 3.1: the origin is a critical point and the remaining orbits are circular periodic orbits centered at the origin.

It follows from (3.3) that each periodic orbit has period $2\pi/f(r)$. Since $f'(r_0) \ne 0$, for $r \ne r_0$ sufficiently close to $r_0$ the corresponding periodic orbit is traversed with angular velocity $f(r) \ne f(r_0)$. This yields the following property. Let $x(t)$ and $x_0(t)$ be solutions of equation (3.3) such that $x(t_0)$ and $x_0(t_0)$ are, respectively, on the circles of radius $r$ and $r_0$. Given $x(t_0)$ arbitrarily close to $x_0(t_0)$, there exists $t > t_0$ such that $x(t)$ and $x_0(t)$ are on the same diameter of the periodic orbits, but on opposite sides of the origin. This shows that the second condition in Definition 3.1 is not satisfied, and thus the solution $x_0(t)$ is unstable.

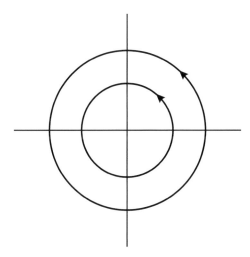

**Figure 3.1.** Phase portrait of equation (3.3).

**3.1.2. Asymptotic stability.** Now we introduce the notion of asymptotic stability for a solution.

**Definition 3.4.** A solution $x(t, t_0, \bar{x}_0)$ of equation (3.1) defined for all $t > t_0$ is said to be *asymptotically stable* if:

   a) $x(t, t_0, \bar{x}_0)$ is stable;
   b) there exists $\alpha > 0$ such that if $\|x_0 - \bar{x}_0\| < \alpha$, then

$$\|x(t, t_0, x_0) - x(t, t_0, \bar{x}_0)\| \to 0 \quad \text{when} \quad t \to +\infty.$$

The following example shows that for a solution to be asymptotically stable it is not sufficient that the second condition in Definition 3.4 is satisfied.

**Example 3.5.** Consider the equation in polar coordinates

$$\begin{cases} r' = r(1 - r), \\ \theta' = \sin^2(\theta/2). \end{cases} \tag{3.4}$$

Its phase portrait is shown in Figure 3.2. We note that the critical point $(1, 0)$ is a solution satisfying the second condition in Definition 3.4 but not the first one (it is sufficient to consider, for example, the solution on the circle of radius 1 centered at the origin).

**Example 3.6.** Consider the equation

$$\begin{cases} r' = r(1 - r), \\ \theta' = \sin^2 \theta. \end{cases} \tag{3.5}$$

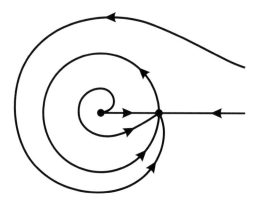

**Figure 3.2.** Phase portrait of equation (3.4).

Its phase portrait is shown in Figure 3.3. We note that since the angular velocity $\theta'$ does not depend on $r$, for any ray $L$ starting at the origin the set

$$L_t = \big\{x(t, t_0, x_0) : x_0 \in L\big\}$$

is still a ray (starting at the origin) for each $t > t_0$. This implies that each solution outside the straight line $y = 0$ is asymptotically stable. On the other hand, all solutions on the straight line $y = 0$ are unstable.

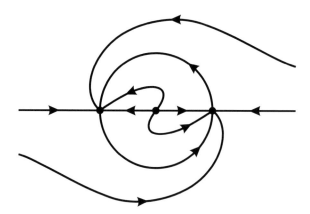

**Figure 3.3.** Phase portrait of equation (3.5).

## 3.2. Stability of linear equations

In this section we consider the particular case of the linear equations

$$x' = A(t)x, \tag{3.6}$$

where $A(t)$ is an $n \times n$ matrix varying continuously with $t \in \mathbb{R}$. After studying the general case, we consider the particular cases of the equations with constant coefficients and periodic coefficients.

**3.2.1. Nonautonomous linear equations: general case.** We first show that in what concerns the study of the stability of linear equations it is sufficient to consider the zero solution.

**Proposition 3.7.** *Let $A \colon \mathbb{R} \to M_n$ be a continuous function. For equation (3.6), the zero solution (with arbitrary initial time $t_0$) is stable (respectively asymptotically stable) if and only if all solutions are stable (respectively asymptotically stable).*

**Proof.** We divide the proof into steps.

*Step 1. Reduction to the zero solution.* Let $X(t)$ be a fundamental solution of equation (3.6). By Theorem 2.8, the solution of the initial value problem (2.2) has maximal interval $\mathbb{R}$ and is given by

$$x(t, t_0, x_0) = X(t)X(t_0)^{-1}x_0. \tag{3.7}$$

It follows from (3.7) that the zero solution (with initial time $t_0$) is stable if and only if given $\varepsilon > 0$, there exists $\delta > 0$ such that

$$\|X(t)X(t_0)^{-1}x_0\| < \varepsilon \quad \text{when} \quad \|x_0\| < \delta \tag{3.8}$$

for $t > t_0$. Since $X(t)X(t_0)^{-1}$ is a linear transformation, this is the same as

$$\|X(t)X(t_0)^{-1}(x_0 - \bar{x}_0)\| < \varepsilon \quad \text{when} \quad \|x_0 - \bar{x}_0\| < \delta,$$

or equivalently,

$$\|x(t, t_0, x_0) - x(t, t_0, \bar{x}_0)\| < \varepsilon \quad \text{when} \quad \|x - \bar{x}_0\| < \delta,$$

for $t > t_0$ and $\bar{x}_0 \in \mathbb{R}^n$. Therefore, the zero solution (with initial time $t_0$) is stable if and only if all solutions (with initial time $t_0$) are stable.

For the asymptotic stability, we note that since $X(t)X(t_0)^{-1}$ is linear, we have

$$\lim_{t \to +\infty} \|X(t)X(t_0)^{-1}x_0\| = 0 \quad \text{when} \quad \|x_0\| < \alpha \tag{3.9}$$

if and only if

$$\lim_{t \to +\infty} \|X(t)X(t_0)^{-1}(x_0 - \bar{x}_0)\| = 0 \quad \text{when} \quad \|x_0 - \bar{x}_0\| < \alpha$$

for every $\bar{x}_0 \in \mathbb{R}^n$. This shows that the zero solution (with initial time $t_0$) is asymptotically stable if and only if all solutions (with initial time $t_0$) are asymptotically stable.

*Step 2. Independence from the initial time.* It remains to verify that the stability and asymptotic stability of the zero solution are independent from the initial time $t_0$. To that effect, take $t_1 \neq t_0$ and note that

$$X(t)X(t_1)^{-1} = X(t)X(t_0)^{-1}X(t_0)X(t_1)^{-1}.$$

It follows from (3.8) that given $\varepsilon > 0$, there exists $\delta > 0$ such that

$$\|X(t)X(t_1)^{-1}x_0\| < \varepsilon \quad \text{when} \quad \|X(t_0)X(t_1)^{-1}x_0\| < \delta,$$

and thus,

$$\|X(t)X(t_1)^{-1}x_0\| < \varepsilon \quad \text{when} \quad \|x_0\| < \frac{\delta}{\|X(t_0)X(t_1)^{-1}\|}, \tag{3.10}$$

for $t > t_0$. If $t_1 \geq t_0$, then property (3.10) holds for $t > t_1$. On the other hand, for $t_1 < t_0$ the function $t \mapsto X(t)X(t_1)^{-1}$ is continuous in $[t_1, t_0]$. Taking

$$\delta \leq \frac{\|X(t_0)X(t_1)^{-1}\|\varepsilon}{\max_{t \in [t_1, t_0]} \|X(t)X(t_1)^{-1}\|},$$

we obtain

$$\|X(t_0)X(t_1)^{-1}x_0\| \leq \max_{t \in [t_1, t_0]} \|X(t)X(t_1)^{-1}\| \cdot \|x_0\| < \delta \leq \varepsilon$$

for $t \in [t_1, t_0]$ and $x_0$ as in (3.10). Therefore, property (3.10) holds for $t > t_1$. This shows that if the zero solution is stable with initial time $t_0$, then it is stable with any initial time.

Now we assume that the zero solution is asymptotically stable with initial time $t_0$. It follows easily from (3.9) that

$$\lim_{t \to +\infty} X(t) = 0,$$

and thus,

$$\lim_{t \to +\infty} \|X(t)X(t_1)^{-1}x_0\| = 0$$

for every $t_1 \in \mathbb{R}$ and $x_0 \in \mathbb{R}^n$. This shows that the zero solution is asymptotically stable with any initial time. $\qquad \square$

It follows from Proposition 3.7 that for equation (3.6) the zero solution (with arbitrary initial time $t_0$) is unstable if and only if all solutions are unstable.

In view of Proposition 3.7 it is natural to introduce the following notion.

**Definition 3.8.** Equation (3.6) is said to be *stable* (respectively, *asymptotically stable* or *unstable*) if all its solutions are stable (respectively, asymptotically stable or unstable).

Now we give a characterization of the notions of stability and asymptotic stability for the nonautonomous linear equation $x' = A(t)x$.

**Theorem 3.9.** *Let $A \colon \mathbb{R} \to M_n$ be a continuous function and let $X(t)$ be a fundamental solution of equation (3.6). Then the equation is:*

a) *stable if and only if* $\sup \left\{ \|X(t)\| : t > 0 \right\} < +\infty$;

b) *asymptotically stable if and only if*

$$\|X(t)\| \to 0 \quad when \quad t \to +\infty.$$

**Proof.** It follows from Proposition 3.7 and (3.8) that the zero solution is stable (with arbitrary initial time $t_0$) if and only if given $\varepsilon > 0$, there exists $\delta > 0$ such that

$$\|X(t)X(0)^{-1}x_0\| \le \varepsilon \quad when \quad \|x_0\| \le \delta,$$

for $t > 0$. Thus, if the zero solution is stable, then

$$
\begin{aligned}
\|X(t)X(0)^{-1}\| &= \sup_{x_0 \ne 0} \frac{\|X(t)X(0)^{-1}x_0\|}{\|x_0\|} \\
&= \sup_{x_0 \ne 0} \frac{\|X(t)X(0)^{-1}(\delta x_0/\|x_0\|)\|}{\|\delta x_0/\|x_0\|\|} \\
&= \sup_{\|y_0\|=\delta} \frac{\|X(t)X(0)^{-1}y_0\|}{\|y_0\|} \le \frac{\varepsilon}{\delta}
\end{aligned}
$$

for $t > 0$, and hence,

$$\sup \left\{ \|X(t)\| : t > 0 \right\} \le \sup \left\{ \|X(t)X(0)^{-1}\| : t > 0 \right\} \|X(0)\|$$
$$\le \frac{\varepsilon \|X(0)\|}{\delta} < +\infty. \tag{3.11}$$

On the other hand, if the supremum in (3.11) is finite, then there exists $C > 0$ such that

$$\|X(t)\| < C \quad for \quad t > 0,$$

and thus,

$$\|X(t)X(0)^{-1}x_0\| < \varepsilon \quad when \quad \|x_0\| < \frac{\varepsilon}{C\|X(0)^{-1}\|},$$

for $t > 0$. This establishes the first property.

For the second property, we first note that if $\|X(t)\| \to 0$ when $t \to +\infty$, then

$$\sup \left\{ \|X(t)\| : t > 0 \right\} < +\infty,$$

and hence, by the first property, the solution is stable. Moreover,

$$\|X(t)X(0)^{-1}x_0\| \to 0 \quad when \quad t \to +\infty, \tag{3.12}$$

for every $x_0 \in \mathbb{R}^n$, and thus, equation (3.6) is asymptotically stable. On the other hand, if the equation is asymptotically stable, then property (3.12) holds for $\|x_0\|$ sufficiently small, and thus $\|X(t)\| \to 0$ when $t \to +\infty$. $\square$

**3.2.2. Constant coefficients and periodic coefficients.** In this section we consider the particular cases of the linear equations with constant coefficients and periodic coefficients. We start with the case of constant coefficients.

**Theorem 3.10.** *For a square matrix $A$, the equation $x' = Ax$ is:*

a) *stable if and only if $A$ has no eigenvalues with positive real part and each eigenvalue with zero real part has a diagonal Jordan block;*

b) *asymptotically stable if and only if $A$ has only eigenvalues with negative real part;*

c) *unstable if and only if $A$ has at least one eigenvalue with positive real part or at least one eigenvalue with zero real part and a nondiagonal Jordan block.*

**Proof.** The claims follow easily from the Jordan canonical form in Theorem 2.16 combined with the description of the exponential of a Jordan block in Proposition 2.18. Indeed, if $S$ is an invertible square matrix satisfying (2.12), then

$$e^{At} = Se^{(S^{-1}AS)t}S^{-1}$$

$$= S \begin{pmatrix} e^{R_1 t} & & 0 \\ & \ddots & \\ 0 & & e^{R_k t} \end{pmatrix} S^{-1},$$

where $R_j$ is the $n_j \times n_j$ matrix in (2.13). On the other hand, it follows from Proposition 2.18 that

$$e^{R_j t} = e^{\lambda_j t}\left(\mathrm{Id} + tN_j + \frac{1}{2!}t^2 N_j^2 + \cdots + \frac{1}{(n_j - 1)!}t^{n_j - 1}N_j^{n_j - 1}\right),$$

where $N_j$ is the $n_j \times n_j$ matrix in (2.15). This shows that each entry of the matrix $e^{R_j t}$ is of the form $e^{\lambda_j t}p(t)$, where $p$ is a polynomial of degree at most $n_j - 1$. The desired result now follows immediately from Theorem 3.9. $\square$

Now we consider linear equations with periodic coefficients.

**Theorem 3.11.** *Let $A \colon \mathbb{R} \to M_n$ be a $T$-periodic continuous function and let $B$ be the matrix in (2.32). Then the equation $x' = A(t)x$ is:*

a) *stable if and only if there are no characteristic exponents with positive real part and for each characteristic exponent with zero real part the corresponding Jordan block of the matrix $B$ is diagonal;*

b) *asymptotically stable if and only if there are only characteristic exponents with negative real part;*

c) *unstable if and only if there is at least one characteristic exponent with positive real part or at least one characteristic exponent with zero real part such that the corresponding Jordan block of the matrix $B$ is not diagonal.*

**Proof.** By Floquet's theorem (Theorem 2.31), any fundamental solution of the equation $x' = A(t)x$ is of the form (2.32). By (2.33), the matrix function $P(t)$ is $T$-periodic, and thus, each property in Theorem 3.9 depends only on the term $e^{Bt}$. That is,

$$\sup\{\|X(t)\| : t > 0\} < +\infty \quad \Leftrightarrow \quad \sup\{\|e^{Bt}\| : t > 0\} < +\infty,$$

and

$$\lim_{t \to +\infty} \|X(t)\| = 0 \quad \Leftrightarrow \quad \lim_{t \to +\infty} \|e^{Bt}\| = 0.$$

This shows that the stability of the equation $x' = A(t)x$ coincides with the stability of the equation $x' = Bx$. Since the eigenvalues of the matrix $B$ are the characteristic exponents (up to an integer multiple of $2\pi i/T$), the desired result follows now immediately from Theorem 3.10. $\square$

### 3.3. Stability under nonlinear perturbations

In this section we consider a class of nonlinear perturbations of an asymptotically stable linear equation. We start with the case of perturbations of linear equations with constant coefficients.

**Theorem 3.12.** *Let $A$ be an $n \times n$ matrix having only eigenvalues with negative real part and let $g \colon \mathbb{R} \times \mathbb{R}^n \to \mathbb{R}^n$ be continuous and locally Lipschitz in $x$. If $g(t,0) = 0$ for every $t \in \mathbb{R}$, and*

$$\limsup_{\substack{x \to 0 \\ t \in \mathbb{R}}} \frac{\|g(t,x)\|}{\|x\|} = 0, \tag{3.13}$$

*then the zero solution of the equation*

$$x' = Ax + g(t,x) \tag{3.14}$$

*is asymptotically stable. Moreover, there exist constants $C, \lambda, \delta > 0$ such that for each $t_0 \in \mathbb{R}$ and each solution $x(t)$ of equation (3.14) with $\|x(t_0)\| < \delta$, we have*

$$\|x(t)\| \le Ce^{-\lambda(t-t_0)}\|x(t_0)\| \quad \text{for} \quad t \ge t_0. \tag{3.15}$$

**Proof.** Since $A$ has only eigenvalues with negative real part, by Proposition 2.27 there exist constants $c, \mu > 0$ such that

$$\|e^{At}\| \le ce^{-\mu t} \tag{3.16}$$

for $t > 0$. On the other hand, by (3.13), given $\varepsilon > 0$, there exists $\delta > 0$ such that

$$\|g(t,x)\| \le \varepsilon\|x\| \tag{3.17}$$

for every $t \in \mathbb{R}$ and $x \in \mathbb{R}^n$ with $\|x\| < \delta$. Now let $x(t)$ be the solution of equation (3.14) with $x(t_0) = x_0$. By the Variation of parameters formula in Theorem 2.26, we have

$$x(t) = e^{A(t-t_0)}x_0 + \int_{t_0}^t e^{A(t-s)}g(s, x(s))\, ds \qquad (3.18)$$

for $t$ in the maximal interval of the solution. Moreover, given $x_0 \in \mathbb{R}^n$ with $\|x_0\| < \delta$, the solution $x(t)$ satisfies $\|x(t)\| < \delta$ for any $t > t_0$ sufficiently close to $t_0$, say for $t \in [t_0, t_1]$. Hence, it follows from (3.16) and (3.17) that

$$\|x(t)\| \le ce^{-\mu(t-t_0)}\|x_0\| + \int_{t_0}^t ce^{-\mu(t-s)}\varepsilon\|x(s)\|\, ds, \qquad (3.19)$$

or equivalently,

$$e^{\mu t}\|x(t)\| \le ce^{\mu t_0}\|x_0\| + \int_{t_0}^t c\varepsilon e^{\mu s}\|x(s)\|\, ds,$$

for $t \in [t_0, t_1]$. By Gronwall's lemma (Proposition 1.39), we obtain

$$e^{\mu t}\|x(t)\| \le ce^{\mu t_0}\|x_0\|e^{c\varepsilon(t-t_0)},$$

that is,

$$\|x(t)\| \le ce^{(-\mu+c\varepsilon)(t-t_0)}\|x_0\|, \qquad (3.20)$$

for $t \in [t_0, t_1]$. Assuming, without loss of generality, that $c \ge 1$ and taking $\varepsilon > 0$ so small that $-\mu + c\varepsilon < 0$, it follows from (3.20) that if $\|x_0\| < \delta/c$, then $\|x(t)\| < \delta$ for $t \in [t_0, t_1]$. Thus, there exists $t_2 > t_1$ such that the solution $x(t)$ is defined in the interval $[t_0, t_2]$ and satisfies $\|x(t)\| < \delta$ for $t \in [t_0, t_2]$. One can repeat this procedure indefinitely (without changing $\varepsilon$ and $\delta$) to conclude that there exists an increasing sequence $t_n$ such that $x(t)$ is defined in the interval $[t_0, t_n]$ and satisfies $\|x(t)\| < \delta$ for $t \in [t_0, t_n]$, for each $n \in \mathbb{N}$.

Now let $b$ be the supremum of all sequences $t_n$. If $b < +\infty$, then we would have $\|x(b^-)\| \le \delta$, which contradicts to Theorem 1.46. Thus, $b = +\infty$ and the solution $x(t)$ is defined in $[t_0, +\infty)$. Moreover, it satisfies (3.20) for all $t \ge t_0$, which yields inequality (3.15). This shows that the zero solution is asymptotically stable. $\qquad \square$

The following is an application of Theorem 3.12.

**Theorem 3.13.** *Let $f: \mathbb{R}^n \to \mathbb{R}^n$ be a function of class $C^1$ and let $x_0 \in \mathbb{R}^n$ be a point with $f(x_0) = 0$ such that $d_{x_0}f$ has only eigenvalues with negative real part. Then there exist constants $C, \lambda, \delta > 0$ such that for each $t_0 \in \mathbb{R}$ the solution $x(t)$ of the initial value problem*

$$\begin{cases} x' = f(x), \\ x(t_0) = \bar{x}_0 \end{cases}$$

*satisfies*

$$\|x(t) - x_0\| \le Ce^{-\lambda(t-t_0)}\|\bar{x}_0 - x_0\| \tag{3.21}$$

*for every $t > t_0$ and $\bar{x}_0 \in \mathbb{R}^n$ with $\|\bar{x}_0 - x_0\| < \delta$.*

**Proof.** We have

$$x' = f(x) = d_{x_0}f(x - x_0) + f(x) - d_{x_0}f(x - x_0). \tag{3.22}$$

Letting $y = x - x_0$, equation (3.22) can be written in the form

$$y' = Ay + g(t, y), \tag{3.23}$$

where $A = d_{x_0}f$ and

$$g(t, y) = f(x_0 + y) - d_{x_0}fy.$$

By hypothesis, the matrix $A$ has only eigenvalues with negative real part. Moreover, $g(t, 0) = f(x_0) = 0$ for every $t \in \mathbb{R}$, and since $f$ is of class $C^1$, we have

$$\sup_{t \in \mathbb{R}} \frac{g(t, y)}{\|y\|} = \frac{f(x_0 + y) - f(x_0) - d_{x_0}fy}{\|y\|} \to 0$$

when $y \to 0$. In other words, the hypotheses of Theorem 3.12 are satisfied. Hence, it follows from (3.15) that there exist constants $C, \lambda, \delta > 0$ such that for each $t_0 \in \mathbb{R}$ and each solution $y(t)$ of equation (3.23) with $\|y(t_0)\| < \delta$, we have

$$\|y(t)\| \le Ce^{-\lambda(t-t_0)}\|y(t_0)\| \quad \text{for} \quad t \ge t_0.$$

This establishes inequality (3.21). $\qquad\square$

One can obtain corresponding results for nonlinear perturbations of nonautonomous linear equations. The following is a version of Theorem 3.12 in this general context.

**Theorem 3.14.** *Let $A\colon \mathbb{R} \to M_n$ be a continuous function and let $X(t)$ be a fundamental solution of the equation $x' = A(t)x$ such that*

$$\|X(t)X(s)^{-1}\| \le ce^{-\mu(t-s)}$$

*for some constants $c, \mu > 0$ and every $t \ge s$. If the function $g\colon \mathbb{R} \times \mathbb{R}^n \to \mathbb{R}$ is continuous and locally Lipschitz in $x$, satisfies $g(t, 0) = 0$ for every $t \in \mathbb{R}$, and property (3.13) holds, then the zero solution of the equation*

$$x' = A(t)x + g(t, x) \tag{3.24}$$

*is asymptotically stable.*

**Proof.** We follow closely the proof of Theorem 3.12, replacing identity (3.18) by an appropriate identity. Namely, if $x(t)$ is the solution of equation (3.24) with $x(t_0) = x_0$, then it follows from the Variation of parameters formula in Theorem 2.25 that

$$x(t) = X(t)X(t_0)^{-1}x_0 + \int_{t_0}^{t} X(t)X(s)^{-1}g(s, x(s))\,ds \qquad (3.25)$$

for $t$ in the maximal interval of the solution. On the other hand, by (3.13), given $\varepsilon > 0$, there exists $\delta > 0$ such that inequality (3.17) holds for every $t \in \mathbb{R}$ and $x \in \mathbb{R}^n$ with $\|x\| < \delta$. Now take $t_1 > t_0$ and $x_0 \in \mathbb{R}^n$ with $\|x_0\| < \delta$ such that $x(t)$ is defined in the interval $[t_0, t_1]$ and satisfies $\|x(t)\| < \delta$ for $t \in [t_0, t_1]$. It follows from (3.25) and (3.17) that inequality (3.19) holds for $t \in [t_0, t_1]$. Now one can repeat the arguments in the proof of Theorem 3.12 to conclude that if $\|x_0\| < \delta/c$, then the solution $x(t)$ is defined in $[t_0, +\infty)$ and satisfies

$$\|x(t)\| \leq c e^{(-\mu + c\varepsilon)(t - t_0)}\|x_0\|$$

for $t \geq t_0$ (assuming that $c \geq 1$ and that $\varepsilon > 0$ is so small that $-\mu + c\varepsilon < 0$). In particular, the zero solution is asymptotically stable. $\qquad\square$

### 3.4. Lyapunov functions

This section is an introduction to the theory of Lyapunov functions, which sometimes allows one to establish the stability or instability of a given solution in a more or less automatic manner.

**3.4.1. Basic notions.** We first recall the notion of locally Lipschitz function.

**Definition 3.15.** A function $f\colon D \to \mathbb{R}^n$ in an open set $D \subset \mathbb{R}^n$ is said to be *locally Lipschitz* if for each compact set $K \subset D$ there exists $L > 0$ such that

$$\|f(x) - f(y)\| \leq L\|x - y\| \qquad (3.26)$$

for every $x, y \in K$.

Let $f\colon D \to \mathbb{R}^n$ be a locally Lipschitz function. One can easily verify that $f$ is locally Lipschitz if and only if the function $g\colon \mathbb{R} \times D \to \mathbb{R}^n$ defined by $g(t, x) = f(x)$ is locally Lipschitz in $x$. Moreover, by (3.26), any locally Lipschitz function is continuous. Now let $\varphi_t(x_0) = x(t, x_0)$ be the solution of the initial value problem

$$\begin{cases} x' = f(x), \\ x(0) = x_0, \end{cases} \qquad (3.27)$$

which in view of the Picard–Lindelöf theorem (Theorem 1.18) is well defined. Given a differentiable function $V\colon D \to \mathbb{R}$, define a new function $\dot{V}\colon D \to \mathbb{R}$ by

$$\dot{V}(x) = \nabla V(x) \cdot f(x).$$

We note that

$$\dot{V}(x) = \left( d_{\varphi_t(x)} V \frac{\partial}{\partial t} \varphi_t(x) \right) \Big|_{t=0} = \frac{\partial}{\partial t} V(\varphi_t(x)) \Big|_{t=0}. \tag{3.28}$$

Now we introduce the notion of a Lyapunov function for a critical point of the equation $x' = f(x)$.

**Definition 3.16.** Given $x_0 \in D$ with $f(x_0) = 0$, a differentiable function $V : D \to \mathbb{R}$ is called a *Lyapunov function* for $x_0$ if there exists an open set $U \subset D$ containing $x_0$ such that:

a) $V(x_0) = 0$ and $V(x) > 0$ for $x \in U \setminus \{x_0\}$;

b) $\dot{V}(x) \le 0$ for $x \in U$.

A Lyapunov function is called a *strict Lyapunov function* if the second condition can be replaced by $\dot{V}(x) < 0$ for $x \in U \setminus \{x_0\}$.

**Example 3.17.** Consider the equation

$$\begin{cases} x' = -x + y, \\ y' = -x - y^3. \end{cases}$$

The origin is the only critical point. We show that the function $V : \mathbb{R}^2 \to \mathbb{R}$ given by

$$V(x, y) = x^2 + y^2$$

is a strict Lyapunov function for $(0, 0)$. We have $V(0, 0) = 0$ and $V(x, y) > 0$ for $(x, y) \ne (0, 0)$. Moreover,

$$\dot{V}(x, y) = (2x, 2y) \cdot (-x + y, -x - y^3)$$
$$= -2(x^2 + y^4) < 0$$

for $(x, y) \ne (0, 0)$.

**3.4.2. Stability criterion.** The existence of a Lyapunov function (respectively, a strict Lyapunov function) for a critical point of a differential equation $x' = f(x)$ allows one to establish the stability (respectively, the asymptotic stability) of that point.

**Theorem 3.18.** *Let $f : D \to \mathbb{R}^n$ be a locally Lipschitz function in an open set $D \subset \mathbb{R}^n$ and let $x_0 \in D$ be a critical point of the equation $x' = f(x)$.*

a) *If there exists a Lyapunov function for $x_0$, then $x_0$ is stable.*

b) *If there exists a strict Lyapunov function for $x_0$, then $x_0$ is asymptotically stable.*

**Proof.** We first assume that there exists a Lyapunov function for $x_0$ in some open set $U \subset D$ containing $x_0$. Take $\varepsilon > 0$ such that $B(x_0, \varepsilon) \subset U$, and

$$m = \min \{V(x) : x \in \partial B(x_0, \varepsilon)\}.$$

Since $V$ is continuous (because it is locally Lipschitz) and $V > 0$ in the set $B(x_0, \varepsilon) \setminus \{x_0\}$, there exists $\delta \in (0, \varepsilon)$ such that

$$0 < \max \{V(x) : x \in \overline{B(x_0, \delta)}\} < m. \tag{3.29}$$

On the other hand, it follows from (3.28) that the function $t \mapsto V(\varphi_t(x))$ is nonincreasing (in the maximal interval $I$ of the solution). Indeed, proceeding as in the proof of Proposition 1.13, we obtain $\varphi_t = \varphi_{t-s} \circ \varphi_s$ for $t$ sufficiently close to $s$, and thus,

$$\frac{\partial}{\partial t} V(\varphi_t(x))\big|_{t=s} = \frac{\partial}{\partial t} V(\varphi_{t-s}(\varphi_s(x)))\big|_{t=s}$$
$$= \dot{V}(\varphi_s(x)) \leq 0 \tag{3.30}$$

for $s \in I$. Hence, it follows from (3.29) that any solution $\varphi_t(x)$ of the initial value problem (3.27) with $x \in \overline{B(x_0, \delta)}$ is contained in $B(x_0, \varepsilon)$ for every $t > 0$ in its maximal interval. This implies that each solution $\varphi_t(x)$ with $x \in \overline{B(x_0, \delta)}$ is defined for all $t > 0$, and thus the critical point $x_0$ is stable.

Now we assume that there exists a strict Lyapunov function for $x_0$. It remains to show that $\varphi_t(x) \to x_0$ when $t \to +\infty$, for any point $x \in B(x_0, \alpha)$ with $\alpha$ sufficiently small. Proceeding as in (3.30), we conclude that for each $x \in U \setminus \{x_0\}$ the function $t \mapsto V(\varphi_t(x))$ is decreasing (in the maximal interval of the solution). Now let $(t_n)_n$ be a sequence of real numbers with $t_n \nearrow +\infty$ such that $(\varphi_{t_n}(x))_n$ converges, and let $y$ be the limit of this sequence. Then

$$V(\varphi_{t_n}(x)) \searrow V(y) \quad \text{when} \quad n \to \infty,$$

because $t \mapsto V(\varphi_t(x))$ is decreasing. Moreover,

$$V(\varphi_t(x)) > V(y) \quad \text{for} \quad t > 0. \tag{3.31}$$

Now we assume that $y \neq x_0$. Then $V(\varphi_s(y)) < V(y)$ for every $s > 0$. Taking $n$ sufficiently large, one can ensure that $\varphi_{t_n+s}(x) = \varphi_s(\varphi_{t_n}(x))$ is as close as desired to $\varphi_s(y)$, and thus also that $V(\varphi_{t_n+s}(x))$ is as close as desired to $V(\varphi_s(y)) < V(y)$. Hence,

$$V(\varphi_{t_n+s}(x)) < V(y),$$

but this contradicts to (3.31). Therefore, $y = x_0$, and we conclude that $\varphi_t(x) \to x_0$ when $t \to +\infty$. $\square$

The following are applications of Theorem 3.18.

**Example 3.19.** Consider the equation

$$\begin{cases} x' = y - xy^2, \\ y' = -x^3. \end{cases} \tag{3.32}$$

This can be seen as a perturbation of the linear equation $(x, y)' = (y, 0)$, which has the phase portrait in Figure 2.11. We consider the critical point $(0, 0)$ of equation (3.32), and the function

$$V(x, y) = x^4 + 2y^2.$$

We have $V(0, 0) = 0$ and $V(x, y) > 0$ for $(x, y) \neq (0, 0)$. Moreover,

$$\dot{V}(x, y) = (4x^3, 4y) \cdot (y - xy^2, -x^3)$$
$$= 4x^3 y - 4x^2 y^2 - 4x^3 y = -4x^2 y^2 \leq 0.$$

Hence, $V$ is a Lyapunov function for $(0, 0)$, and it follows from Theorem 3.18 that the origin is stable.

**Example 3.20.** Consider the equation

$$\begin{cases} x' = x^2 - x - y, \\ y' = x. \end{cases} \tag{3.33}$$

We discuss the stability of the critical point at the origin. Equation (3.33) can be written in the form

$$\begin{pmatrix} x \\ y \end{pmatrix} = \begin{pmatrix} -1 & -1 \\ 1 & 0 \end{pmatrix} \begin{pmatrix} x \\ y \end{pmatrix} + \begin{pmatrix} x^2 \\ 0 \end{pmatrix}. \tag{3.34}$$

Since the $2 \times 2$ matrix in (3.34) has eigenvalues $(-1 \pm i\sqrt{3})/2$, both with negative real part, it follows from Theorem 3.13 that the origin is asymptotically stable.

One can also use Lyapunov functions to study the stability of the origin. However, it is not always easy to find a strict Lyapunov function (when it exists). For example, consider the function $V(x, y) = x^2 + y^2$. We have

$$\dot{V}(x, y) = (2x, 2y) \cdot (x^2 - x - y, x) = 2x^2(x - 1),$$

and thus, $\dot{V}(x, y) \leq 0$ in a sufficiently small neighborhood of $(0, 0)$. Hence, $V$ is a Lyapunov function, and it follows from Theorem 3.18 that the origin is stable. However, this does not show that the origin is asymptotically stable. To that effect, consider the function

$$W(x, y) = x^2 + xy + y^2.$$

We have $W(0, 0) = 0$. Moreover,

$$W(x, y) = \begin{pmatrix} x \\ y \end{pmatrix}^* \begin{pmatrix} 1 & 1/2 \\ 1/2 & 1 \end{pmatrix} \begin{pmatrix} x \\ y \end{pmatrix}. \tag{3.35}$$

Since the $2 \times 2$ matrix in (3.35) has eigenvalues $1/2$ and $3/2$, the matrix is positive definite. In particular, $W(x,y) > 0$ for $(x,y) \neq 0$. We also have

$$
\begin{aligned}
\dot{W}(x,y) &= (2x + y, x + 2y) \cdot (x^2 - x - y, x) \\
&= 2x^3 - x^2 + x^2 y - xy - y^2 \\
&= -\begin{pmatrix} x \\ y \end{pmatrix}^* \begin{pmatrix} 1 & 1/2 \\ 1/2 & 1 \end{pmatrix} \begin{pmatrix} x \\ y \end{pmatrix} + 2x^3 + x^2 y.
\end{aligned}
\tag{3.36}
$$

Since the $2 \times 2$ matrix in (3.36) is positive definite, there exist constants $a, b > 0$ such that

$$
-a\|(x,y)\|^2 \le -\begin{pmatrix} x \\ y \end{pmatrix}^* \begin{pmatrix} 1 & 1/2 \\ 1/2 & 1 \end{pmatrix} \begin{pmatrix} x \\ y \end{pmatrix} \le -b\|(x,y)\|^2
$$

for every $(x,y) \in \mathbb{R}^2$. Moreover,

$$
\frac{2x^3 + x^2 y}{\|(x,y)\|^2} \to 0 \quad \text{when} \quad (x,y) \to 0.
$$

Therefore, given $\varepsilon > 0$, we have

$$
-a - \varepsilon < \frac{\dot{W}(x,y)}{\|(x,y)\|^2} < -b + \varepsilon
$$

for any sufficiently small $(x,y) \neq 0$. Taking $\varepsilon$ so small that $-b + \varepsilon < 0$, we obtain $\dot{W}(x,y) < 0$ for any sufficiently small $(x,y) \neq 0$. Hence, it follows from Theorem 3.18 that the origin is asymptotically stable.

**Example 3.21.** Consider the equation

$$
x'' + f(x) = 0,
\tag{3.37}
$$

where $f \colon \mathbb{R} \to \mathbb{R}$ is a function of class $C^1$ with $f(0) = 0$ such that

$$
x f(x) > 0 \quad \text{for} \quad x \neq 0
\tag{3.38}
$$

(that is, $f(x)$ and $x$ always have the same sign). This corresponds to apply a force $-f(x)$ to a particle of mass 1 that is moving without friction. Condition (3.38) corresponds to assume that the force always points to the origin.

Equation (3.37) can be written in the form

$$
\begin{cases} x' = y, \\ y' = -f(x), \end{cases}
\tag{3.39}
$$

and $(0,0)$ is a critical point. We use a Lyapunov function to show that the origin is stable. Namely, consider the function

$$
V(x,y) = \frac{1}{2}y^2 + \int_0^x f(s)\,ds,
\tag{3.40}
$$

which corresponds to the sum of the kinetic energy $y^2/2$ (recall that the particle has mass 1) with the potential energy $\int_0^x f(s)\,ds$. We have $V(0,0) = 0$ and $V(x,y) > 0$ for $(x,y) \neq (0,0)$, due to condition (3.38). Moreover,

$$\dot{V}(x,y) = (f(x), y) \cdot (y, -f(x)) = 0,$$

and thus, $V$ is a Lyapunov function for $(0,0)$. Hence, it follows from Theorem 3.18 that the origin is stable. Incidentally, along the solutions we have

$$\frac{d}{dt}V(x,y) = yy' + f(x)x'$$
$$= -yf(x) + f(x)y = 0,$$

and thus, equation (3.39) is conservative. This corresponds to the conservation of energy.

**Example 3.22.** Given $\varepsilon > 0$, consider the equation

$$x'' + \varepsilon x' + f(x) = 0, \tag{3.41}$$

with $f \colon \mathbb{R} \to \mathbb{R}$ as in Example 3.21. Equation (3.41) can be written in the form

$$\begin{cases} x' = y, \\ y' = -f(x) - \varepsilon y. \end{cases}$$

We consider again the function $V$ in (3.40), which satisfies $V(0,0) = 0$ and $V(x,y) > 0$ for $(x,y) \neq (0,0)$. Moreover,

$$\dot{V}(x,y) = (f(x), y) \cdot (y, -f(x) - \varepsilon y) = -\varepsilon y^2 \leq 0,$$

and $V$ is a Lyapunov function for $(0,0)$. Hence, it follows from Theorem 3.18 that the origin is stable.

**3.4.3. Instability criterion.** We conclude this chapter with the description of an instability criterion for the critical points of an equation $x' = f(x)$. The criterion is analogous to the stability criterion in Theorem 3.18.

**Theorem 3.23.** *Let $f \colon D \to \mathbb{R}^n$ be a function of class $C^1$ in an open set $D \subset \mathbb{R}^n$ and let $x_0 \in D$ be a critical point of the equation $x' = f(x)$. Also, let $V \colon U \to \mathbb{R}$ be a function of class $C^1$ in a neighborhood $U \subset D$ of $x_0$ such that:*

    a) *$V(x_0) = 0$ and $\dot{V}(x) > 0$ for $x \in U \setminus \{x_0\}$;*

    b) *$V$ takes positive values in any neighborhood of $x_0$.*

*Then the critical point $x_0$ is unstable.*

**Proof.** Let $A \subset U$ be a neighborhood of $x_0$ and let $\varphi_t(x)$ be the solution of the initial value problem (3.27). If in each neighborhood $A$ there exists a solution $\varphi_t(x)$ that is not defined for all $t > 0$, then there is nothing to show.

Hence, one can assume that all solutions $\varphi_t(x)$ with $x \in A$ are defined for all $t > 0$.

Now take $y \in A$ with $V(y) > 0$ (which by hypothesis always exists). Since $\dot{V}(x) > 0$ for $x \in U \setminus \{x_0\}$, proceeding as in (3.30) we conclude that the function $t \mapsto V(\varphi_t(y))$ is increasing whenever $\varphi_t(y) \in U$. Thus, since $V(x_0) = 0$, the solution $\varphi_t(y)$ does not come near $x_0$, that is, there exists a neighborhood $B$ of $x_0$ such that $\varphi_t(y) \notin B$ for $t > 0$. Now we assume that the solution does not leave $A$ and we define

$$m = \inf \{ \dot{V}(\varphi_t(y)) : t > 0 \}.$$

Since $\dot{V} = \nabla V \cdot f$ is continuous and $\overline{A \setminus B}$ is compact, we have

$$m \geq \inf \{ \dot{V}(x) : x \in \overline{A \setminus B} \} > 0$$

(because continuous functions with values in $\mathbb{R}$ have a minimum in each compact set). We also have

$$V(\varphi_t(y)) \geq V(y) + mt \quad \text{for} \quad t > 0.$$

Thus, there exists $T > 0$ such that

$$V(\varphi_T(y)) > \max \{ V(x) : x \in \overline{A} \},$$

and hence $\varphi_T(y) \notin A$. This contradiction shows that there exist points $x$ arbitrarily close to $x_0$ (because by hypothesis $V$ takes positive values in any neighborhood of $x_0$) such that the solution $\varphi_t(x)$ leaves the neighborhood $A$. Therefore, the critical point $x_0$ is unstable. $\qquad\square$

The following is an application of Theorem 3.23.

**Example 3.24.** Consider the equation

$$\begin{cases} x' = 3x + y^3, \\ y' = -2y + x^2, \end{cases}$$

for which $(0, 0)$ is a critical point, and the function

$$V(x, y) = x^2 - y^2.$$

Clearly, $V(0, 0) = 0$ and $V$ takes positive values in any neighborhood of $(0, 0)$. Moreover,

$$\dot{V}(x, y) = (2x, -2y) \cdot (3x + y^3, -2y + x^2)$$
$$= 6x^2 + 4y^2 + 2xy^3 - 2x^2 y.$$

Since

$$\frac{\dot{V}(x, y)}{6x^2 + 4y^2} \to 1 \quad \text{when} \quad (x, y) \to (0, 0),$$

we have $\dot{V}(x, y) > 0$ for any sufficiently small $(x, y) \neq (0, 0)$. Hence, it follows from Theorem 3.23 that the origin is unstable.

## 3.5. Exercises

**Exercise 3.1.** Find all stable, asymptotically stable and unstable solutions of the equation:

    a) $x' = x(x-2)$;

    b) $x'' + 4x = 0$.

**Exercise 3.2.** For a function $g \colon \mathbb{R}^n \to \mathbb{R}$ of class $C^2$, consider the equation

$$x' = \nabla g(x).$$

    a) Show that if $u$ is a nonconstant solution, then $g \circ u$ is strictly increasing.

    b) Show that there are no periodic orbits.

    c) Determine the stability of the origin when $g(x, y) = x^2 + y^4$.

    d) Determine the stability of the origin when $g(x, y) = x^2 y^4$.

**Exercise 3.3.** Consider the equation in polar coordinates

$$\begin{cases} r' = f(r), \\ \theta' = 1, \end{cases}$$

where

$$f(r) = \begin{cases} r\sin(1/r^2), & r \neq 0, \\ 0, & r = 0. \end{cases}$$

Show that the origin is stable but is not asymptotically stable.

**Exercise 3.4.** Determine the stability of the zero solution of the equation:

    a) $\begin{cases} x' = -x + xy, \\ y' = x - y - x^2 - y^3; \end{cases}$

    b) $\begin{cases} x' = -x + x^2 + y^2, \\ y' = 2x - 3y + y^3; \end{cases}$

    c) $\begin{cases} x' = -x + 2x(x + y)^2, \\ y' = -y^3 + 2y^3(x + y)^2; \end{cases}$

    d) $\begin{cases} x' = x^3 - 3xy^2, \\ y' = 3x^2y - y^3. \end{cases}$

**Exercise 3.5.** Let $a \colon \mathbb{R} \to \mathbb{R}$ be a $T$-periodic continuous function.

    a) Find the characteristic exponent of the equation $x' = a(t)x$.

    b) Find a necessary and sufficient condition in terms of the function $a$ so that the zero solution is asymptotically stable.

**Exercise 3.6.** Compute

$$\lambda(x) = \limsup_{t \to +\infty} \frac{1}{t} \log \|x(t)\|$$

(with the convention that $\log 0 = -\infty$) for each solution $x(t)$ of the equation:

a) $x'' + x = 0$;

b) $x' = [a + b(\sin \log t + \cos \log t)]x$ with $a, b \in \mathbb{R}$.

**Exercise 3.7.** Given an $n \times n$ matrix $A$, define $\chi \colon \mathbb{C}^n \to [-\infty, +\infty)$ by

$$\chi(v) = \limsup_{n \to +\infty} \frac{1}{n} \log \|A^n v\|.$$

Show that:

a) $\chi(\alpha v) = \chi(v)$ for $\alpha \neq 0$;

b) $\chi(v + w) \leq \max\{\chi(v), \chi(w)\}$;

c) if $\chi(v) \neq \chi(w)$, then $\chi(v + w) = \max\{\chi(v), \chi(w)\}$;

d) $\chi$ takes only finitely many values.

**Exercise 3.8.** Show that if $(x_1(t), x_2(t))$ is a nonzero solution of the equation

$$\begin{cases} x_1' = [-1.01 - (\sin \log t + \cos \log t)]x_1, \\ x_2' = [-1.01 + (\sin \log t + \cos \log t)]x_2, \end{cases}$$

then

$$\limsup_{t \to +\infty} \frac{1}{t} \log \|(x_1(t), x_2(t))\| < 0.$$

**Exercise 3.9.** Consider the equation

$$\begin{cases} y_1' = [-1.01 - (\sin \log t + \cos \log t)]y_1, \\ y_2' = [-1.01 + (\sin \log t + \cos \log t)]y_2 + y_1{}^2. \end{cases}$$

a) Verify that each solution $(y_1(t), y_2(t))$, with $t > 0$, can be written in the form

$$\begin{cases} y_1(t) = c_1 e^{-1.01t - a(t)}, \\ y_2(t) = c_2 e^{-1.01t + a(t)} + c_1{}^2 e^{-1.01t + a(t)} \int_s^t e^{-3a(\tau) - 1.01\tau} \, d\tau \end{cases}$$

for some constants $c_1$, $c_2$ and $s$, where $a(t) = t \sin \log t$.

b) Taking $\varepsilon \in (0, \pi/4)$ and defining $t_k = e^{2k\pi - \pi/2}$ for $k \in \mathbb{N}$, verify that

$$-3a(\tau) \geq 3\tau \cos \varepsilon \quad \text{for} \quad \tau \in [t_k e^{-\varepsilon}, t_k],$$

and conclude that

$$\int_s^{t_k} e^{-3a(\tau) - 1.01\tau} \, d\tau \geq \int_{t_k e^{-\varepsilon}}^{t_k} e^{-3a(\tau) - 1.01\tau} \, d\tau \geq c e^{(3\cos \varepsilon - 1.01)t_k}$$

for any sufficiently large $k \in \mathbb{N}$, where $c > 0$ is a constant.

c) Show that there exists a solution $(y_1(t), y_2(t))$ such that

$$\limsup_{t \to +\infty} \frac{1}{t} \log \|(y_1(t), y_2(t))\| > 0.$$

Hint: Consider the times $t = t_k e^\pi$.

**Exercise 3.10.** Let $f \colon \mathbb{R}^n \to \mathbb{R}^n$ be a function of class $C^\infty$ such that the equation $x' = f(x)$ defines a flow $\varphi_t$ in $\mathbb{R}^n$.

a) Show that

$$\varphi_t(x) = x + f(x)t + \frac{1}{2}(d_x f)f(x)t^2 + o(t^2).$$

b) Verify that

$$\det d_x \varphi_t = 1 + \operatorname{div} f(x)t + o(t).$$

c) Given an open set $A \subset \mathbb{R}^n$ and $t \in \mathbb{R}$, show that

$$\frac{d}{dt}\mu(\varphi_t(A)) = \int_{\varphi_t(A)} \operatorname{div} f,$$

where $\mu$ denotes the volume in $\mathbb{R}^n$.

d) Show that if $\operatorname{div} f = 0$, then the equation $x' = f(x)$ has neither asymptotically stable critical points nor asymptotically stable periodic solutions.

**Solutions.**

**3.1** a) The solutions in $(-\infty, 2)$ are asymptotically stable and the solutions in $[2, +\infty)$ are unstable.
    b) All solutions are stable but none are asymptotically stable.

**3.2** c) Unstable.
    d) Unstable.

**3.4** a) Asymptotically stable.
    b) Asymptotically stable.
    c) Asymptotically stable.
    d) Unstable.

**3.5** a) $(1/T) \int_0^T a(s)\, ds$.
    b) $\int_0^T a(s)\, ds < 0$.

**3.6** a) $\lambda(x) = 0$.
    b) $\lambda(x) = a + |b|$.

# Hyperbolicity and Topological Conjugacies

This chapter is dedicated to the study of hyperbolicity and its consequences, particularly at the level of stability. After a brief introduction to the notion of hyperbolicity, we establish a fundamental result on the behavior of the solutions in a neighborhood of a hyperbolic critical point—the Grobman–Hartman theorem. It shows that the solutions of a sufficiently small perturbation of an equation with a hyperbolic critical point are topologically conjugate to the solutions of its linear variational equation. We also show that the topological conjugacy is Hölder continuous. For additional topics we refer the reader to [**7, 15, 19, 23**].

## 4.1. Hyperbolic critical points

In this section we introduce the notion of hyperbolic critical point.

We recall that a square matrix is said to be hyperbolic if all its eigenvalues have nonzero real part (see Definition 2.48). Now let $f\colon \mathbb{R}^n \to \mathbb{R}^n$ be a function of class $C^1$.

**Definition 4.1.** A point $x_0 \in \mathbb{R}^n$ with $f(x_0) = 0$ such that the matrix $d_{x_0} f$ is hyperbolic is called a *hyperbolic critical point* of the equation $x' = f(x)$.

Given a hyperbolic critical point $x_0 \in \mathbb{R}^n$ of $x' = f(x)$, consider the linear equation

$$x' = Ax, \quad \text{where} \quad A = d_{x_0} f.$$

We recall that its solutions are given by

$$x(t) = e^{A(t-t_0)} x(t_0), \quad t \in \mathbb{R}.$$

**Definition 4.2.** Given a hyperbolic critical point $x_0 \in \mathbb{R}^n$ of the equation $x' = f(x)$, we define the *stable* and *unstable spaces* of $x_0$, respectively, by

$$E^s = \left\{ x \in \mathbb{R}^n : e^{At} x \to 0 \text{ when } t \to +\infty \right\}$$

and

$$E^u = \left\{ x \in \mathbb{R}^n : e^{At} x \to 0 \text{ when } t \to -\infty \right\}.$$

**Proposition 4.3.** *If $x_0 \in \mathbb{R}^n$ is a hyperbolic critical point of the equation $x' = f(x)$, then:*

    a) *$E^s$ and $E^u$ are linear subspaces of $\mathbb{R}^n$ with $E^s \oplus E^u = \mathbb{R}^n$;*

    b) *for every $x \in E^s$, $y \in E^u$ and $t \in \mathbb{R}$, we have*

$$e^{At} x \in E^s \quad \text{and} \quad e^{At} y \in E^u.$$

**Proof.** Since the matrix $A = d_{x_0} f$ has no eigenvalues with zero real part, its Jordan canonical form can be written in the form

$$\begin{pmatrix} A_s & 0 \\ 0 & A_u \end{pmatrix}$$

with respect to some decomposition $\mathbb{R}^n = F^s \oplus F^u$, where $A_s$ and $A_u$ correspond respectively to the Jordan blocks of eigenvalues with negative real part and the Jordan blocks of eigenvalues with positive real part. It follows from Proposition 2.27 that

$$e^{At} x \to 0 \quad \text{when} \quad t \to +\infty,$$

for $x \in F^s$, and that

$$e^{At} x \to 0 \quad \text{when} \quad t \to -\infty,$$

for $x \in F^u$. Hence,

$$F^s = E^s \quad \text{and} \quad F^u = E^u,$$

which establishes the first property.

For the second property, we first recall that

$$e^{A\tau} e^{At} = e^{At} e^{A\tau}$$

for every $t, \tau \in \mathbb{R}$ (see Exercise 2.3). In particular, if $x \in E^s$ and $t \in \mathbb{R}$, then

$$e^{A\tau} \left( e^{At} x \right) = e^{At} \left( e^{A\tau} x \right). \tag{4.1}$$

Since $e^{A\tau} x \to 0$ when $\tau \to +\infty$, we obtain

$$e^{At} \left( e^{A\tau} x \right) \to 0 \quad \text{when} \quad \tau \to +\infty,$$

and it follows from (4.1) that $e^{At} x \in E^s$. One can show in an analogous manner that if $y \in E^u$ and $t \in \mathbb{R}$, then $e^{At} y \in E^u$. $\square$

By Proposition 4.3, the spaces $E^s$ and $E^u$ associated to a hyperbolic critical point form a direct sum. Hence, given $x \in \mathbb{R}^n$, there exist unique points $y \in E^s$ and $z \in E^u$ such that $x = y + z$. We define $P_s, P_u \colon \mathbb{R}^n \to \mathbb{R}^n$ by

$$P_s x = y \quad \text{and} \quad P_u x = z. \tag{4.2}$$

One can easily verify that $P_s$ and $P_u$ are linear transformations with

$$P_s(\mathbb{R}^n) = E^s \quad \text{and} \quad P_u(\mathbb{R}^n) = E^u.$$

Moreover, $P_s^2 = P_s$ and $P_u^2 = P_u$, that is, $P_s$ and $P_u$ are projections, respectively, over the spaces $E^s$ and $E^u$. One can also show that

$$E^s = \operatorname{Re} G^s + \operatorname{Im} G^s \quad \text{and} \quad E^u = \operatorname{Re} G^u + \operatorname{Im} G^u,$$

where $G^s$ and $G^u$ are the subspaces of $\mathbb{C}^n$ generated by the root spaces, respectively, of eigenvalues with negative real part and eigenvalues with positive real part. More precisely,

$$G^s = \left\{ x \in \mathbb{C}^n : (d_{x_0} f - \lambda \operatorname{Id})^k x = 0 \text{ for some } k \in \mathbb{N}, \lambda \in \mathbb{C} \text{ with } \operatorname{Re} \lambda < 0 \right\}$$

and

$$G^u = \left\{ x \in \mathbb{C}^n : (d_{x_0} f - \lambda \operatorname{Id})^k x = 0 \text{ for some } k \in \mathbb{N}, \lambda \in \mathbb{C} \text{ with } \operatorname{Re} \lambda > 0 \right\}.$$

## 4.2. The Grobman–Hartman theorem

Let $f \colon \mathbb{R}^n \to \mathbb{R}^n$ be a function of class $C^1$ such that the equation $x' = f(x)$ has a hyperbolic critical point $x_0$. In this section we show that the solutions of the equations

$$x' = f(x) \quad \text{and} \quad y' = d_{x_0} f y$$

are topologically conjugate, respectively, in neighborhoods of $x_0$ and $0$. More precisely, and in an analogous manner to that in Definition 2.44, this means that if $\psi_t(z)$ and $\varphi_t(z)$ are, respectively, the solutions of the initial value problems

$$\begin{cases} x' = f(x), \\ x(0) = z \end{cases} \quad \text{and} \quad \begin{cases} y' = d_{x_0} f y, \\ y(0) = z, \end{cases} \tag{4.3}$$

then there exists a homeomorphism $h \colon U \to V$, where $U$ and $V$ are, respectively, neighborhoods of $x_0$ and $0$, such that $h(x_0) = 0$ and

$$h(\psi_t(z)) = \varphi_t(h(z)) \tag{4.4}$$

whenever $z, \psi_t(z) \in U$. This guarantees that the phase portraits of the equations in (4.3) are homeomorphic, respectively, in neighborhoods of $x_0$ and $0$ (see Figure 4.1).

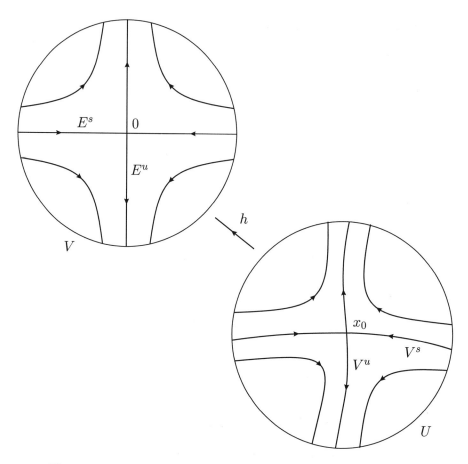

**Figure 4.1.** Topological conjugacy between the solutions of the equations in (4.3).

### 4.2.1. Perturbations of hyperbolic matrices.

We first establish a result on the existence of topological conjugacies for the perturbations of a linear equation $x' = Ax$ with a hyperbolic matrix $A$.

**Theorem 4.4.** *Let $A$ be an $n \times n$ hyperbolic matrix and let $g \colon \mathbb{R}^n \to \mathbb{R}^n$ be a bounded function such that $g(0) = 0$ and*

$$\|g(x) - g(y)\| \le \delta \|x - y\| \tag{4.5}$$

*for every $x, y \in \mathbb{R}^n$. For any sufficiently small $\delta$, there exists a unique bounded continuous function $\eta \colon \mathbb{R}^n \to \mathbb{R}^n$ such that $\eta(0) = 0$ and*

$$h \circ e^{At} = \psi_t \circ h, \quad t \in \mathbb{R}, \tag{4.6}$$

*where $h = \mathrm{Id} + \eta$ and $\psi_t$ is the flow determined by the equation*

$$x' = Ax + g(x). \tag{4.7}$$

*Moreover, $h$ is a homeomorphism.*

**Proof.** We divide the proof into several steps.

*Step 1. Existence of global solutions.* We first show that equation (4.7) defines a flow. Each solution $x(t)$ satisfies

$$x(t) = e^{A(t-t_0)}x(t_0) + \int_{t_0}^t e^{A(t-s)}g(x(s))\,ds \tag{4.8}$$

for $t$ in the corresponding maximal interval $I_x$ (we note that the function $(t,x) \mapsto Ax + g(x)$ is continuous and locally Lipschitz in $x$, and thus one can apply Theorem 1.43). It follows from (4.5) with $y = 0$ and (4.8) that

$$\|x(t)\| \leq e^{\|A\|(t-t_0)}\|x(t_0)\| + \int_{t_0}^t e^{\|A\|(t-s)}\delta\|x(s)\|\,ds$$

for $t \geq t_0$ in $I_x$. By Gronwall's lemma (Proposition 1.39), we obtain

$$e^{-\|A\|t}\|x(t)\| \leq e^{-\|A\|t_0}\|x(t_0)\|e^{\delta(t-t_0)},$$

or equivalently,

$$\|x(t)\| \leq e^{(\|A\|+\delta)(t-t_0)}\|x(t_0)\|, \tag{4.9}$$

for $t \geq t_0$ in $I_x$. This implies that each solution $x(t)$ is defined in $[t_0, +\infty)$. Otherwise, the right endpoint $b$ of the interval $I_x$ would be finite, and by (4.9) we would have

$$\|x(b^-)\| \leq e^{(\|A+\delta\|)(b-t_0)}\|x(t_0)\|,$$

which contradicts to Theorem 1.46 on the behavior of solutions at the endpoints of the maximal interval. One can show in a similar manner that all solutions are defined in the interval $(-\infty, t_0]$.

*Step 2. Construction of an auxiliary function.* Let $X_0$ be the set of bounded continuous functions $\eta\colon \mathbb{R}^n \to \mathbb{R}^n$ with $\eta(0) = 0$. It follows from Proposition 1.30 that $X_0$ is a complete metric space with the distance

$$d(\eta, \xi) = \sup\left\{\|\eta(x) - \xi(x)\| : x \in \mathbb{R}^n\right\}. \tag{4.10}$$

We define a transformation $G$ in $X_0$ by

$$G(\eta)(x) = \int_0^{+\infty} P_s e^{A\tau}\eta(e^{-A\tau}x)\,d\tau - \int_0^{+\infty} P_u e^{-A\tau}\eta(e^{A\tau}x)\,d\tau$$
$$= \int_{-\infty}^t P_s e^{A(t-\tau)}\eta(e^{A(\tau-t)}x)\,d\tau - \int_t^{+\infty} P_u e^{A(t-\tau)}\eta(e^{A(\tau-t)}x)\,d\tau, \tag{4.11}$$

for each $\eta \in X_0$ and $x \in \mathbb{R}^n$, where $P_s$ and $P_u$ are, respectively, the projections over the stable and unstable spaces (see (4.2)). We first show that the transformation $G$ is well defined. Since the matrix $A$ is hyperbolic, by Proposition 2.27 there exist constants $c, \mu > 0$ such that

$$\|P_s e^{A\tau}\| \leq ce^{-\mu\tau} \quad \text{and} \quad \|P_u e^{-A\tau}\| \leq ce^{-\mu\tau} \tag{4.12}$$

for $\tau \geq 0$. Thus,

$$\|P_s e^{A\tau} \eta(e^{-A\tau})\| \leq c e^{-\mu\tau} \|\eta\|_\infty \tag{4.13}$$

and

$$\|P_u e^{-A\tau} \eta(e^{A\tau})\| \leq c e^{-\mu\tau} \|\eta\|_\infty \tag{4.14}$$

for $\tau \geq 0$, where

$$\|\eta\|_\infty := \sup \left\{ \|\eta(x)\| : x \in \mathbb{R}^n \right\}.$$

Since $\eta \in X_0$, we have $\|\eta\|_\infty < +\infty$. Hence, it follows from (4.13) and (4.14) that

$$\int_0^{+\infty} \|P_s e^{A\tau} \eta(e^{-A\tau}x)\| \, d\tau + \int_0^{+\infty} \|P_u e^{-A\tau} \eta(e^{A\tau}x)\| \, d\tau$$
$$\leq \int_0^{+\infty} c e^{-\mu\tau} \|\eta\|_\infty \, d\tau + \int_0^{+\infty} c e^{-\mu\tau} \|\eta\|_\infty \, d\tau = \frac{2c}{\mu} \|\eta\|_\infty, \tag{4.15}$$

and the transformation $G$ is well defined. Moreover, by (4.15), we have

$$\|G(\eta)(x)\| \leq \frac{2c}{\mu} \|\eta\|_\infty$$

for each $x \in \mathbb{R}^n$, and the function $G(\eta)$ is bounded for each $\eta \in X_0$. Now we show that $G(\eta)$ is continuous. Given $x \in \mathbb{R}^n$ and $\varepsilon > 0$, it follows from (4.15) that there exists $T > 0$ such that

$$\int_T^{+\infty} \|P_s e^{A\tau} \eta(e^{-A\tau}x)\| \, d\tau + \int_T^{+\infty} \|P_u e^{-A\tau} \eta(e^{A\tau}x)\| \, d\tau < \varepsilon.$$

For each $x, y \in \mathbb{R}^n$, we have

$$\|G(\eta)(x) - G(\eta)(y)\| < 2\varepsilon + \int_0^T \left\| P_s e^{A\tau} [\eta(e^{-A\tau}x) - \eta(e^{-A\tau}y)] \right\| \, d\tau$$
$$+ \int_0^T \left\| P_u e^{-A\tau} [\eta(e^{A\tau}x) - \eta(e^{A\tau}y)] \right\| \, d\tau$$
$$\leq 2\varepsilon + \int_0^T c e^{-\mu\tau} \|\eta(e^{-A\tau}x) - \eta(e^{-A\tau}y)\| \, d\tau \tag{4.16}$$
$$+ \int_0^T c e^{-\mu\tau} \|\eta(e^{A\tau}x) - \eta(e^{A\tau}y)\| \, d\tau.$$

For each $\delta > 0$, the function $(t, x) \mapsto \eta(e^{-At}x)$ is uniformly continuous on the set $[0, T] \times \overline{B(x, \delta)}$ (because continuous functions are uniformly continuous on each compact set). Hence, there exists $\delta > 0$ such that

$$\|\eta(e^{-A\tau}x) - \eta(e^{-A\tau}y)\| < \varepsilon$$

for $\tau \in [0, T]$ and $y \in \overline{B(x, \delta)}$. Analogously, there exists $\delta > 0$ (which one can assume to be the same as before) such that

$$\|\eta(e^{A\tau}x) - \eta(e^{A\tau}y)\| < \varepsilon$$

for $\tau \in [0, T]$ and $y \in \overline{B(x, \delta)}$. It follows from (4.16) that

$$\|G(\eta)(x) - G(\eta)(y)\| < 2\varepsilon + 2cT\varepsilon$$

for $y \in \overline{B(x, \delta)}$. This shows that the function $G(\eta)$ is continuous. Moreover, by (4.11), we have $G(\eta)(0) = 0$ and thus $G(X_0) \subset X_0$.

*Step 3. Equivalent form of identity* (4.6). Given $\eta \in X_0$, let

$$g_\eta(x) = g(h(x)),$$

where $h = \operatorname{Id} + \eta$. We note that the function $g_\eta$ is bounded, because $g$ is bounded. It follows from (4.5) that

$$\|g_\eta(x) - g_\eta(y)\| \leq \delta\|x - y\| + \delta\|\eta(x) - \eta(y)\|.$$

Moreover,

$$g_\eta(0) = g(h(0)) = g(0) = 0,$$

and thus $g_\eta \in X_0$.

**Lemma 4.5.** *Identity* (4.6) *is equivalent to* $G(g_\eta) = \eta$.

**Proof of the lemma.** If $G(g_\eta) = \eta$, then it follows from Leibniz's rule that

$$\frac{\partial}{\partial t} e^{-At} h(e^{At} x)$$

$$= \frac{\partial}{\partial s} e^{-A(t+s)} \eta(e^{A(t+s)} x)\big|_{s=0}$$

$$= e^{-At} \frac{\partial}{\partial s} e^{-As} \eta(e^{As} e^{At} x)\big|_{s=0}$$

$$= e^{-At} \frac{\partial}{\partial s} e^{-As} G(g_\eta)(e^{As} x)\big|_{s=0}$$

$$= e^{-At} \frac{\partial}{\partial s} \left( \int_{-\infty}^{s} P_s e^{-A\tau} g_\eta(e^{A\tau} x) \, d\tau - \int_{s}^{+\infty} P_u e^{-A\tau} g_\eta(e^{A\tau} x) \, d\tau \right)\big|_{s=0}$$

$$= e^{-At} \left( P_s g_\eta(x) + P_u g_\eta(x) \right)$$

$$= e^{-At} g(h(e^{At} x)). \tag{4.17}$$

On the other hand,

$$\frac{\partial}{\partial t} e^{-At} h(e^{At} x) = -A e^{-At} h(e^{At} x) + e^{-At} \frac{\partial}{\partial t} h(e^{At} x),$$

and it follows from (4.17) that

$$-A h(e^{At} x) + \frac{\partial}{\partial t} h(e^{At} x) = g(h(e^{At} x)),$$

that is,

$$\frac{\partial}{\partial t} h(e^{At} x) = A h(e^{At} x) + g(h(e^{At} x)).$$

This shows that the function $y(t) = h(e^{At}x)$ satisfies equation (4.7). Since $y(0) = h(x)$, we obtain

$$h(e^{At}x) = \psi_t(h(x))$$

for every $x \in \mathbb{R}^n$, which yields identity (4.6).

Now we assume that (4.6) holds. Since $h = \mathrm{Id} + \eta$, one can rewrite this identity in the form

$$\mathrm{Id} + \eta = \psi_t \circ h \circ e^{-At}. \tag{4.18}$$

On the other hand, by the Variation of parameters formula in Theorem 2.26, we have

$$\psi_t(x) = e^{At}x + \int_0^t e^{A(t-\tau)}g(\psi_\tau(x))\,d\tau. \tag{4.19}$$

Using (4.6) and (4.19), it follows from (4.18) that

$$\begin{aligned}
\eta(x) &= \psi_t(h(e^{-At}x)) - x \\
&= e^{At}h(e^{-At}x) + \int_0^t e^{A(t-\tau)}g(\psi_\tau(h(e^{-At}x)))\,d\tau - x \\
&= e^{At}\eta(e^{-At}x) + \int_0^t e^{A(t-\tau)}g(h(e^{A(\tau-t)}x))\,d\tau \\
&= e^{At}\eta(e^{-At}x) + \int_0^t e^{A\tau}g(h(e^{-A\tau}x))\,d\tau.
\end{aligned} \tag{4.20}$$

Moreover, it follows from (4.13) that

$$\|P_s e^{At}\eta(e^{-At}x)\| \le ce^{-\mu t}\|\eta\|_\infty \to 0$$

when $t \to +\infty$, and thus, by (4.20),

$$P_s\eta(x) = \int_0^{+\infty} P_s e^{A\tau}g(h(e^{-A\tau}x))\,d\tau. \tag{4.21}$$

On the other hand, it follows from the third equality in (4.20) that

$$e^{-At}\eta(x) = \eta(e^{-At}x) + \int_0^t e^{-A\tau}g(h(e^{A(\tau-t)}x))\,d\tau.$$

Substituting $x$ by $e^{At}x$, we obtain

$$e^{-At}\eta(e^{At}x) = \eta(x) + \int_0^t e^{-A\tau}g(h(e^{A\tau}x))\,d\tau. \tag{4.22}$$

Analogously, it follows from (4.14) that

$$\|P_u e^{-At}\eta(e^{At}x)\| \le ce^{-\mu t}\|\eta\|_\infty \to 0$$

when $t \to -\infty$, and thus, by (4.22),

$$P_u\eta(x) = -\int_0^{+\infty} P_u e^{-A\tau}g(h(e^{A\tau}x))\,d\tau. \tag{4.23}$$

Adding (4.21) and (4.23), we finally obtain

$$\eta(x) = P_s\eta(x) + P_u\eta(x)$$

$$= \int_0^{+\infty} P_s e^{A\tau} g(h(e^{-A\tau}x)) \, d\tau - \int_0^{+\infty} P_u e^{-A\tau} g(h(e^{A\tau}x)) \, d\tau$$

$$= G(g \circ h)(x) = G(g_\eta)(x).$$

This yields the desired result. $\square$

*Step 4. Construction of the conjugacy.* Now we show that the equation $G(g_\eta) = \eta$ has a unique solution $\eta$.

**Lemma 4.6.** *For any sufficiently small $\delta$, there exists a unique function $\eta \in X_0$ satisfying $G(g_\eta) = \eta$.*

**Proof of the lemma.** We define a transformation $F$ in $X_0$ by

$$F(\eta) = G(g_\eta). \tag{4.24}$$

Since $g_\eta \in X_0$ for $\eta \in X_0$, we have $F(X_0) \subset X_0$ (because $G(X_0) \subset X_0$). Now we show that $F$ is a contraction. For each $\eta, \xi \in X_0$, we have

$$F(\eta)(x) - F(\xi)(x) = \int_{-\infty}^t P_s e^{A(t-\tau)} \big[g_\eta(e^{A(\tau-t)}x) - g_\xi(e^{A(\tau-t)}x)\big] \, d\tau$$

$$- \int_t^{+\infty} P_u e^{A(t-\tau)} \big[g_\eta(e^{A(\tau-t)}x) - g_\xi(e^{A(\tau-t)}x)\big] \, d\tau.$$

It follows from (4.5) that

$$\|g_\eta(e^{A(\tau-t)}x) - g_\xi(e^{A(\tau-t)}x)\| \le \delta\|\eta(e^{A(\tau-t)}x) - \xi(e^{A(\tau-t)}x)\|$$

$$\le \delta d(\eta, \xi),$$

and hence, using (4.12), we obtain

$$\|F(\eta)(x) - F(\xi)(x)\| \le \int_{-\infty}^t c e^{-\mu(t-\tau)} \delta d(\eta, \xi) \, d\tau$$

$$+ \int_t^{+\infty} c e^{-\mu(t-\tau)} \delta d(\eta, \xi) \, d\tau$$

$$= \frac{2c\delta}{\mu} d(\eta, \xi).$$

Thus,

$$d(F(\eta), F(\xi)) \le \frac{2c\delta}{\mu} d(\eta, \xi)$$

and $F$ is a contraction for any sufficiently small $\delta$. Since $X_0$ is a complete metric space and $F(X_0) \subset X_0$, it follows from Theorem 1.35 that there exists a unique $\eta \in X_0$ such that $F(\eta) = \eta$. $\square$

By Lemma 4.5, the function $\eta$ given by Lemma 4.6 satisfies identity (4.6). It remains to show that $h = \mathrm{Id} + \eta$ is a homeomorphism.

*Step 5. Existence of the inverse.* We first show that $h$ is surjective. Given $y \in \mathbb{R}^n$, the equation $h(x) = y$ has a solution of the form $x = y + z$ if and only if

$$z = -\eta(y + z). \tag{4.25}$$

Now we recall Brouwer's fixed point theorem: any continuous function $H \colon B \to B$ in a closed ball $B \subset \mathbb{R}^n$ has at least one fixed point (for a proof see, for example, [10]). Applying this theorem to the function

$$z \mapsto H(z) = -\eta(y + z)$$

in the closed ball $B = \overline{B(0, \|\eta\|_\infty)}$, we conclude that there exists at least one point $z \in B$ satisfying (4.25). This shows that $h$ is surjective.

Now we show that $h$ is injective. Given $x, y \in \mathbb{R}^n$ such that $h(x) = h(y)$, it follows from (4.6) that

$$h(e^{At}x) = h(e^{At}y)$$

for $t \in \mathbb{R}$, and thus,

$$e^{At}(x - y) = -\eta(e^{At}x) + \eta(e^{At}y). \tag{4.26}$$

If $P_s(x - y) \neq 0$ or $P_u(x - y) \neq 0$, then the left-hand side of (4.26) is unbounded while the right-hand side is bounded. This contradiction shows that

$$P_s(x - y) = P_u(x - y) = 0$$

and hence $x = y$.

*Step 6. Continuity of the inverse.* We first recall the Invariance of domain theorem: if $f \colon U \to \mathbb{R}^n$ is an injective continuous function in an open set $U \subset \mathbb{R}^n$, then $V = f(U)$ is open and $f|U \colon U \to V$ is a homeomorphism (for a proof see, for example, [10]). Since the function $h \colon \mathbb{R}^n \to \mathbb{R}^n$ is continuous and bijective, it follows from the theorem that $h$ is a homeomorphism.  $\square$

**4.2.2. Hyperbolic critical points.** Now we establish the result described at the beginning of Section 4.2 as a consequence of Theorem 4.4.

**Theorem 4.7** (Grobman–Hartman theorem). *Let $f \colon \mathbb{R}^n \to \mathbb{R}^n$ be a function of class $C^1$ and let $x_0 \in \mathbb{R}^n$ be a hyperbolic critical point of the equation $x' = f(x)$. If $\psi_t(z)$ and $\varphi_t(z)$ are, respectively, the solutions of the initial value problems in (4.3), then there exists a homeomorphism $h \colon U \to V$, where $U$ and $V = h(U)$ are, respectively, neighborhoods of $x_0$ and $0$, such that $h(x_0) = 0$ and $h(\psi_t(z)) = \varphi_t(h(z))$ whenever $z, \psi_t(z) \in U$.*

**Proof.** Without loss of generality, we assume that $x_0 = 0$. Indeed, making the change of variables $y = x - x_0$, the equation $x' = f(x)$ becomes $y' = F(y)$, where $F(y) = f(x_0 + y)$. Moreover, $F(0) = f(x_0) = 0$.

Taking $x_0 = 0$, we write

$$f(x) = Ax + g(x),$$

where

$$A = d_0 f \quad \text{and} \quad g(x) = f(x) - Ax.$$

We note that the matrix $A$ is hyperbolic and that the function $g$ is of class $C^1$. Moreover, $g(0) = 0$. In order to apply Theorem 4.4 it remains to verify property (4.5). However, in general, this property may fail, which leads us to modify the function $g$ outside a neighborhood of 0. More precisely, given $\delta > 0$ as in Theorem 4.4, take $r > 0$ so small that

$$\sup \left\{ \|d_x g\| : x \in B(0, r) \right\} \le \delta/3 \tag{4.27}$$

(this is always possible, because the function $x \mapsto d_x g$ is continuous and $d_0 g = 0$). Since $g(0) = 0$, it follows from the Mean value theorem that

$$\sup \left\{ \|g(x)\| : x \in B(0, r) \right\} \le \delta r/3. \tag{4.28}$$

Now we consider a function $\alpha \colon \mathbb{R}^n \to [0, 1]$ of class $C^1$ such that:

 a) $\alpha(x) = 1$ for $x \in B(0, r/3)$;
 b) $\alpha(x) = 0$ for $x \in \mathbb{R}^n \setminus B(0, r)$;
 c) $\sup \left\{ \|d_x \alpha\| : x \in \mathbb{R}^n \right\} \le 2/r$.

For the function $\bar{g} \colon \mathbb{R}^n \to \mathbb{R}^n$ of class $C^1$ defined by

$$\bar{g}(x) = \alpha(x) g(x), \tag{4.29}$$

we have $\bar{g}(0) = 0$. Moreover, by (4.27) and (4.28), we obtain

$$\|d_x \bar{g}\| = \|d_x \alpha g(x) + \alpha(x) d_x g\|$$
$$\le \sup_{x \in \mathbb{R}^n} \|d_x \alpha\| \sup_{x \in B(0,r)} \|g(x)\| + \sup_{x \in \mathbb{R}^n} \|d_x g\| \tag{4.30}$$
$$< \frac{2}{r} \cdot \frac{\delta r}{3} + \frac{\delta}{3} = \delta.$$

Hence, it follows from the Mean value theorem that property (4.5) holds for the function $\bar{g}$. Thus, one can apply Theorem 4.4 to obtain a homeomorphism $h = \mathrm{Id} + \eta$ satisfying

$$h \circ e^{At} = \bar{\psi}_t \circ h, \quad t \in \mathbb{R},$$

where $\bar{\psi}_t(z)$ is the solution of the initial value problem

$$\begin{cases} x' = Ax + \bar{g}(x), \\ g(0) = z. \end{cases}$$

But since $\bar{g}$ coincides with $g$ on the ball $B(0, r/3)$, we have $\psi_t(z) = \bar{\psi}_t(z)$ whenever $\|z\| < r/3$ and $\|\psi_t(z)\| < r/3$. This establishes identity (4.4) for any $t \in \mathbb{R}$ and $z \in \mathbb{R}^n$ such that $\|z\| < r/3$ and $\|\psi_t(z)\| < r/3$. $\qquad\square$

**4.2.3. Stability under perturbations.** It should be noted that Theorems 4.4 and 4.7 do not require the stable and unstable spaces to have positive dimension, that is, the theorems also include the case when one of them is the whole space $\mathbb{R}^n$. This observation has the following consequence.

**Theorem 4.8.** *Let $A$ be an $n \times n$ hyperbolic matrix and let $g\colon \mathbb{R}^n \to \mathbb{R}^n$ be a bounded function with $g(0) = 0$ such that property (4.5) holds for every $x, y \in \mathbb{R}^n$.*

  a) *If $A$ has only eigenvalues with negative real part and $\delta$ is sufficiently small, then the zero solution of equation (4.7) is asymptotically stable.*

  b) *If $A$ has at least one eigenvalue with positive real part and $\delta$ is sufficiently small, then the zero solution of equation (4.7) is unstable.*

**Proof.** It follows from (4.6) that

$$\psi_t = h \circ e^{At} \circ h^{-1}, \tag{4.31}$$

where $h$ is the homeomorphism given by Theorem 4.4. Since $A$ has only eigenvalues with negative real part, we have $\|e^{At}\| \to 0$ when $t \to +\infty$. Thus, since $h(0) = 0$, it follows from (4.31) that the origin is asymptotically stable. Analogously, when $A$ has at least one eigenvalue with positive real part, we have $\|e^{At}\| \to +\infty$ when $t \to +\infty$, and it follows from (4.31) that the origin is unstable. $\qquad\square$

The following result is a consequence of Theorem 4.7.

**Theorem 4.9.** *Let $f\colon \mathbb{R}^n \to \mathbb{R}^n$ be a function of class $C^1$ and let $x_0 \in \mathbb{R}^n$ be a hyperbolic critical point of the equation $x' = f(x)$.*

  a) *If the matrix $d_{x_0} f$ has only eigenvalues with negative real part, then the critical point $x_0$ is asymptotically stable.*

  b) *If the matrix $d_{x_0} f$ has at least one eigenvalue with positive real part, then the critical point $x_0$ is unstable.*

**Proof.** It is sufficient to consider identity (4.4) and proceed in a similar manner to that in the proof of Theorem 4.8. $\qquad\square$

We note that the first property in Theorem 4.9 is also a consequence of Theorem 3.13.

## 4.3. Hölder conjugacies

It is natural to ask whether the homeomorphisms in Theorems 4.4 and 4.7 have more regularity. Unfortunately, in general, it is not possible to obtain conjugacies of class $C^1$, although a detailed discussion falls outside the scope of the book. It is, however, possible to show that those homeomorphisms are always Hölder continuous. More precisely, we have the following result.

**Theorem 4.10.** *Let $A$ be a hyperbolic $n \times n$ matrix and let $g \colon \mathbb{R}^n \to \mathbb{R}^n$ be a bounded function with $g(0) = 0$ such that property (4.5) holds. Given a sufficiently small $\alpha \in (0,1)$ and $K > 0$, there exists $\delta > 0$ such that the homeomorphism $h$ in Theorem 4.4 satisfies*

$$\|h(x) - h(y)\| \leq K\|x - y\|^\alpha$$

*for every $x, y \in \mathbb{R}^n$ with $\|x - y\| < 1$.*

**Proof.** Let $X_0$ be the set of bounded continuous functions $\eta \colon \mathbb{R}^n \to \mathbb{R}^n$ with $\eta(0) = 0$. Given $\alpha \in (0,1)$ and $K > 0$, we consider the subset $X_\alpha \subset X_0$ composed of the functions $\eta \in X_0$ such that

$$\|\eta(x) - \eta(y)\| \leq K\|x - y\|^\alpha$$

for every $x, y \in \mathbb{R}^n$ with $\|x - y\| < 1$. Using Proposition 1.30, one can easily verify that $X_\alpha$ is a complete metric space with the distance in (4.10). Thus, in order to prove the theorem it suffices to show that the transformation $F$ in (4.24) satisfies $F(X_\alpha) \subset X_\alpha$. Indeed, we already know from the proof of Theorem 4.4 that $F$ is a contraction in $X_0$.

Take $\eta \in X_\alpha$. We have

$$\|h(x) - h(y)\| = \|x - y + \eta(x) - \eta(y)\|$$
$$\leq \|x - y\| + \|\eta(x) - \eta(y)\|.$$

Since $\eta \in X_\alpha$, whenever $\|x - y\| < 1$ we obtain

$$\|h(x) - h(y)\| \leq \|x - y\| + K\|x - y\|^\alpha$$
$$\leq (1 + K)\|x - y\|^\alpha. \tag{4.32}$$

Now we consider the difference

$$W = F(\eta)(x) - F(\eta)(y)$$
$$= \int_{-\infty}^{t} P_s e^{A(t-\tau)} \left[ g_\eta(e^{A(\tau-t)}x) - g_\eta(e^{A(\tau-t)}y) \right] d\tau$$
$$- \int_{t}^{+\infty} P_u e^{A(t-\tau)} \left[ g_\eta(e^{A(\tau-t)}x) - g_\eta(e^{A(\tau-t)}y) \right] d\tau,$$

where $g_\eta = g \circ h$. From the proof of Theorem 4.4, we know that $g_\eta \in X_0$. Thus, it follows from (4.5) and (4.12) that

$$\|W\| \le c \int_{-\infty}^{t} e^{-\mu(t-\tau)} \delta \|\eta(e^{A(\tau-t)}x) - \eta(e^{A(\tau-t)}y)\| \, d\tau$$

$$+ c \int_{t}^{+\infty} e^{-\mu(t-\tau)} \delta \|\eta(e^{A(\tau-t)}x) - \eta(e^{A(\tau-t)}y)\| \, d\tau.$$

On the other hand, by (2.11), we have $\|e^{At}\| \le e^{\|A\|\cdot|t|}$ for $t \in \mathbb{R}$. Hence, by (4.32), we obtain

$$\|W\| \le c\delta(1+K) \int_{-\infty}^{t} e^{-\mu(t-\tau)} \|e^{A(\tau-t)}(x-y)\|^\alpha \, d\tau$$

$$+ c\delta(1+K) \int_{t}^{+\infty} e^{-\mu(t-\tau)} \|e^{A(\tau-t)}(x-y)\|^\alpha \, d\tau$$

$$\le c\delta(1+K)\|x-y\|^\alpha \int_{-\infty}^{t} e^{-\mu(t-\tau)} e^{\alpha\|A\|(t-\tau)} \, d\tau$$

$$+ c\delta(1+K)\|x-y\|^\alpha \int_{t}^{+\infty} e^{-\mu(\tau-t)} e^{\alpha\|A\|(\tau-t)} \, d\tau$$

$$= c\delta(1+K)\|x-y\|^\alpha \int_{-\infty}^{t} e^{-(\mu-\alpha\|A\|)(t-\tau)} \, d\tau$$

$$+ c\delta(1+K)\|x-y\|^\alpha \int_{t}^{+\infty} e^{-(\mu-\alpha\|A\|)(\tau-t)} \, d\tau.$$

For any sufficiently small $\alpha$, we have $\mu - \alpha\|A\| > 0$, and hence,

$$\|W\| \le \frac{2c\delta(1+K)}{\mu - \alpha\|A\|} \|x-y\|^\alpha. \tag{4.33}$$

Thus, for any sufficiently small $\delta$, it follows from (4.33) that

$$\|F(\eta)(x) - F(\eta)(y)\| \le K\|x-y\|^\alpha$$

for every $x, y \in \mathbb{R}^n$ with $\|x-y\| < 1$. This completes the proof of the theorem. $\qquad \square$

Under the assumptions of Theorem 4.10 one can also show that

$$\|h^{-1}(x) - h^{-1}(y)\| \le K\|x-y\|^\alpha$$

for every $x, y \in \mathbb{R}^n$ with $\|x-y\| < 1$, eventually making $\alpha$ and $\delta$ sufficiently smaller.

Now we consider the particular case of a hyperbolic critical point. The following result is a simple consequence of Theorems 4.7 and 4.10.

**Theorem 4.11.** *Let $f \colon \mathbb{R}^n \to \mathbb{R}^n$ be a function of class $C^1$ and let $x_0 \in \mathbb{R}^n$ be a hyperbolic critical point of the equation $x' = f(x)$. If $\psi_t(z)$ and $\varphi_t(z)$*

*are, respectively, the solutions of the initial value problems in* (4.3), *then for the homeomorphism h in Theorem 4.7, given a sufficiently small* $\alpha \in (0, 1)$ *and* $K > 0$, *there exists* $d > 0$ *such that*

$$\|h(x) - h(y)\| \le K\|x - y\|^{\alpha}$$

*for every* $x, y \in U$ *with* $\|x - y\| < d$.

## 4.4. Structural stability

We also establish a result on the existence of topological conjugacies for arbitrary linear and nonlinear perturbations.

**Theorem 4.12** (Structural stability). *Let* $f\colon \mathbb{R}^n \to \mathbb{R}^n$ *be a function of class* $C^1$ *and let* $x_0$ *be a hyperbolic critical point of the equation* $x' = f(x)$. *If* $g\colon \mathbb{R}^n \to \mathbb{R}^n$ *is a function of class* $C^1$ *and*

$$a := \sup_{x \in \mathbb{R}^n} \|f(x) - g(x)\| + \sup_{x \in \mathbb{R}^n} \|d_x f - d_x g\|$$

*is sufficiently small, then the solutions of the equations*

$$x' = f(x) \quad and \quad x' = g(x)$$

*are topologically conjugate, respectively, in neighborhoods of* $x_0$ *and* $x_g$, *where* $x_g$ *is the unique critical point of* $x' = g(x)$ *in a sufficiently small neighborhood of* $x_0$.

**Proof.** We first show that the equation $x' = g(x)$ has a unique critical point $x_g$ in a sufficiently small neighborhood of $x_0$, provided that $a$ is sufficiently small. To that effect, consider the function $H\colon C^1(\mathbb{R}^n) \times \mathbb{R}^n \to \mathbb{R}^n$ defined by

$$H(g, y) = g(y),$$

where $C^1(\mathbb{R}^n)$ is the set of all functions $g\colon \mathbb{R}^n \to \mathbb{R}^n$ of class $C^1$ such that

$$\|g\|_{C^1} := \sup_{x \in \mathbb{R}^n} \|g(x)\| + \sup_{x \in \mathbb{R}^n} \|d_x g\| < +\infty.$$

One can show that the Implicit function theorem holds in the space $C^1(\mathbb{R}^n)$ when equipped with the norm $\|\cdot\|_{C^1}$. Now we show that $H$ is of class $C^1$, taking the norm

$$\|(g, y)\| = \|g\|_{C^1} + \|y\|$$

in $C^1(\mathbb{R}^n) \times \mathbb{R}^n$. For each $(g, y), (h, z) \in C^1(\mathbb{R}^n) \times \mathbb{R}^n$ and $\varepsilon \in \mathbb{R}$, we have

$$\frac{H((g, y) + \varepsilon(h, z)) - H(g, y)}{\varepsilon} = \frac{(g + \varepsilon h)(y + \varepsilon z) - g(y)}{\varepsilon}$$

$$= \frac{g(y + \varepsilon z) - g(y)}{\varepsilon} + h(y + \varepsilon z)$$

$$\to d_y g z + h(y)$$

when $\varepsilon \to 0$. Take

$$A := H((g, y) + (h, z)) - H(g, y) - h(y) - d_y g z$$
$$= (g + h)(y + z) - g(y) - h(y) - d_y g z$$
$$= [g(y + z) - g(y) - d_y g z] + [h(y + z) - h(y)].$$

Since

$$h(y + z) - h(y) = \int_0^1 d_{y+tz} h z \, dt,$$

we have

$$\|h(y + z) - h(y)\| \le \|h\|_{C^1} \|z\|,$$

and thus,

$$\frac{\|A\|}{\|(h, z)\|} \le \frac{\|g(y + z) - g(y) - d_y g z\|}{\|z\|} + \frac{\|h\|_{C^1} \|z\|}{\|h\|_{C^1} + \|z\|} \to 0$$

when $\|(h, z)\| \to 0$. This shows that the function $H$ has a derivative, given by

$$d_{(g,y)} H(h, z) = h(y) + d_y g z. \tag{4.34}$$

In order to verify that $H$ is of class $C^1$, we note that

$$\|(d_{(g,y)} H - d_{(\bar{g}, \bar{y})} H)(h, z)\|$$
$$\le \|h(y) - h(\bar{y})\| + \|(d_y g - d_{\bar{y}} \bar{g}) z\|$$
$$\le \|h\|_{C^1} \|y - \bar{y}\| + \|d_y g - d_y \bar{g}\| \cdot \|z\| + \|d_y \bar{g} - d_{\bar{y}} \bar{g}\| \cdot \|z\|$$
$$\le \|h\|_{C^1} \|y - \bar{y}\| + \|g - \bar{g}\|_{C^1} \|z\| + \|d_y \bar{g} - d_{\bar{y}} \bar{g}\| \cdot \|z\|,$$

and thus,

$$\|d_{(g,y)} H - d_{(\bar{g}, \bar{y})} H\| \le \|y - \bar{y}\| + \|g - \bar{g}\|_{C^1} + \|d_y \bar{g} - d_{\bar{y}} \bar{g}\|. \tag{4.35}$$

Since $d\bar{g}$ is continuous, it follows readily from (4.35) that the function $dH$ is also continuous. Moreover, by (4.34), we have

$$\frac{\partial H}{\partial y}(g, y) z = d_{(g,y)} H(0, z) = d_y g z,$$

and thus,

$$\frac{\partial H}{\partial y}(f, x_0) = d_{x_0} f.$$

Since the matrix $d_{x_0} f$ is invertible (because it is hyperbolic), it follows from the Implicit function theorem that for $g \in C^1(\mathbb{R}^n)$ with $\|f - g\|_{C^1}$ sufficiently small, there exists a unique $x_g \in \mathbb{R}^n$ in a sufficiently small neighborhood of $x_0$ such that $g(x_g) = 0$. Moreover, the function $g \mapsto x_g$ is continuous.

Now we consider the pairs of equations

$$x' = f(x) \quad \text{and} \quad y' = d_{x_0} f y \tag{4.36}$$

and

$$x' = g(x) \quad \text{and} \quad y' = d_{x_g} g y. \tag{4.37}$$

By the Grobman–Hartman theorem (Theorem 4.7) there exist topological conjugacies between the solutions of the equations in (4.36) respectively in neighborhoods of the points $x_0$ and 0, as well as between the solutions of the equations in (4.37) respectively in neighborhoods of the points $x_g$ and 0. To complete the proof, it is sufficient to observe that since the eigenvalues of a matrix vary continuously with its entries and the function $g \mapsto x_g$ is continuous, for $\|f - g\|_{C^1}$ sufficiently small the matrix $d_{x_g}g$ is also hyperbolic and

$$m(d_{x_g}g) = m(d_{x_0}f),$$

where $m(A)$ is the number of eigenvalues of the square matrix $A$ with positive real part, counted with their multiplicities. Hence, it follows from Theorem 2.50 that the solutions of the equations $y' = d_{x_0}fy$ and $y' = d_{x_g}gy$ are topologically conjugate. Since the composition of topological conjugacies is still a topological conjugacy, we obtain the desired result. □

## 4.5. Exercises

**Exercise 4.1.** Find all hyperbolic critical points of the equation defined by:

a) $f(x) = x(x - 1)$;

b) $f(x) = x^3$;

c) $f(x, y) = (x + y, y - x^2)$.

**Exercise 4.2.** Determine the spaces $E^s$ and $E^u$ of the origin for the function:

a) $f(x, y, z, w) = \begin{pmatrix} 1 & -3 & 0 & 0 \\ 3 & 1 & 0 & 0 \\ 0 & 0 & -2 & 1 \\ 0 & 0 & -1 & -2 \end{pmatrix} \begin{pmatrix} x \\ y \\ z \\ w \end{pmatrix}$;

b) $f(x, y) = (x - y^2, x^2 - y)$.

**Exercise 4.3.** Under the assumptions of Theorem 4.4, show that if a transformation $h$ satisfying (4.6) is differentiable, then:

a) $d_{e^{At}x}he^{At} = d_{h(x)}\psi_t d_x h$;

b) $d_{e^{At}x}hAe^{At}x = Ah(e^{At}x) + g(h(e^{At}x))$;

c) $Be^{At} = d_0\psi_t B$, where $B = d_0 h$;

d) $d_x hAx = Ah(x) + g(h(x))$.

**Exercise 4.4.** Verify that $h(x, y) = (x, y + x^4/5)$ is a differentiable conjugacy between the solutions of the equations

$$\begin{cases} x' = x, \\ y' = -y \end{cases} \quad \text{and} \quad \begin{cases} x' = x, \\ y' = -y + x^3. \end{cases}$$

**Exercise 4.5.** Discuss the stability of the zero solution of the equation

$$\begin{cases} x' = x - 2\sin x + y, \\ y' = y + xy. \end{cases}$$

**Exercise 4.6.** Consider the equation

$$\begin{cases} x' = yf(x,y), \\ y' = -xf(x,y), \end{cases} \tag{4.38}$$

where $f(x,y) = (x^2 + y^2 - 1)(y - x^2 - 2)$.

a) Sketch the phase portrait.

b) Show that all solutions are global.

c) Find all homoclinic orbits.

d) Find all periodic orbits.

e) Show that in a neighborhood of the origin the periods $T(r)$ of the periodic orbits, where $r = \sqrt{x^2 + y^2}$, satisfy $T(r) = \pi + ar^2 + o(r^2)$ for some $a \neq 0$.

f) Find whether the solutions of equation (4.38) are topologically conjugate to the solutions of the linear equation

$$\begin{cases} x' = yf(0,0), \\ y' = -xf(0,0) \end{cases}$$

in a neighborhood of the origin.

**Exercise 4.7.** Consider the equation

$$\begin{cases} x' = yg(x,y), \\ y' = -xg(x,y), \end{cases}$$

where

$$g(x,y) = (x^2 + y^2 - 1)[y^2 - (x^2 + 2)^2][x^2 - (y^2 + 2)^2].$$

a) Sketch the phase portrait.

b) Find all global solutions.

c) Find all homoclinic orbits.

d) Find all periodic orbits.

e) Show that in a neighborhood of the origin the periods $T(r)$ of the periodic orbits satisfy $T'(r) \neq 0$ for some $r \neq 0$.

**Exercise 4.8.** Consider the equation $x'' + \sin x = 0$.

a) Find an integral for the equation.

b) Sketch the phase portrait.

   c) Determine the stability of all critical points.

   d) Find whether there exist periodic orbits of arbitrarily large period.

   e) Find whether there exist periodic orbits of arbitrarily small period.

   f) Find whether there exist periodic orbits of period $\pi^4$.

**Solutions.**

**4.1** a) 0 and 1.

   b) There are no critical points.

   c) $(0,0)$ and $(-1,1)$.

**4.2** a) $E^s = \{(0,0)\} \times \mathbb{R}^2$ and $E^u = \mathbb{R}^2 \times \{(0,0)\}$.

   b) $E^s = \{0\} \times \mathbb{R}$ and $E^u = \mathbb{R} \times \{0\}$.

**4.5** It is unstable.

**4.6** c) $\{(x,y) \in \mathbb{R}^2 : x^2 + y^2 = 4\} \setminus \{(0,2)\}$.

   d) $\{(x,y) \in \mathbb{R}^2 : x^2 + y^2 = r^2\}$ with $r \in (0,1) \cup (1,2)$.

   f) They are not.

**4.7** b) All solutions are global.

   c) There are no homoclinic orbits.

   d) $\{(x,y) \in \mathbb{R}^2 : x^2 + y^2 = r^2\}$ with $r \in (0,1) \cup (1,2)$.

**4.8** a) $y^2/2 - \cos x$.

   c) $((2n+1)\pi, 0)$, $n \in \mathbb{Z}$ is unstable and $(2n\pi, 0)$, $n \in \mathbb{Z}$ is stable but not asymptotically stable.

   d) Yes.

   e) No.

   f) Yes.

LIVERPOOL JOHN MOORES UNIVERSITY
LEARNING SERVICES

# Chapter 5

# Existence of Invariant Manifolds

In this chapter we continue the study of the consequences of hyperbolicity. In particular, we establish the Hadamard–Perron theorem on the existence of invariant manifolds tangent to the stable and unstable spaces of a hyperbolic critical point. The proofs use the fixed point theorem for a contraction in a complete metric space (Theorem 1.35) and the Fiber contraction theorem (Theorem 1.38). For additional topics we refer the reader to [**7, 18, 19, 23**].

## 5.1. Basic notions

Let $f\colon \mathbb{R}^n \to \mathbb{R}^n$ be a function of class $C^1$ and let $x_0$ be a hyperbolic critical point of the equation $x' = f(x)$. Also, let $\psi_t(z)$ and $\varphi_t(z)$ be, respectively, the solutions of the initial value problems in (4.3). By the Grobman–Hartman theorem (Theorem 4.7), there exists a homeomorphism $h\colon U \to V$, where $U$ and $V$ are, respectively, neighborhoods of $x_0$ and $0$, such that $h(x_0) = 0$ and the identity

$$h(\psi_t(z)) = \varphi_t(h(z)) \tag{5.1}$$

holds whenever $z, \psi_t(z) \in U$. In particular, $h^{-1}$ transforms each orbit $\varphi_t(h(z))$ of the linear equation $y' = Ay$, where $A = d_{x_0}f$, into the orbit $\psi_t(z)$ of the equation $x' = f(x)$. This yields the qualitative behavior in Figure 4.1. Moreover, the stable and unstable spaces $E^s$ and $E^u$ (see Definition 4.2) are transformed by the homeomorphism $h^{-1}$ onto the sets

$$V^s = h^{-1}(E^s \cap V) \quad \text{and} \quad V^u = h^{-1}(E^u \cap V). \tag{5.2}$$

Without loss of generality, one can always assume that the open set $V$ satisfies

$$\varphi_t(E^s \cap V) \subset E^s \cap V \quad \text{and} \quad \varphi_{-t}(E^u \cap V) \subset E^u \cap V \tag{5.3}$$

for every $t > 0$. Indeed, using the construction in the proof of Theorem 2.50, one can show that there exist open sets $B^s \subset E^s$ and $B^u \subset E^u$ containing the origin such that

$$\varphi_t(B^s) \subset B^s \quad \text{and} \quad \varphi_{-t}(B^u) \subset B^u$$

for every $t > 0$. Namely, given $r > 0$, one can take

$$B^s = \left\{ x \in E^s : \int_0^{+\infty} \|e^{At}x\|^2 \, dt < r \right\}$$

and

$$B^u = \left\{ x \in E^u : \int_{-\infty}^0 \|e^{At}x\|^2 \, dt < r \right\}.$$

Then the open set $V = B^s \times B^u$ satisfies (5.3) for every $t > 0$. Moreover, taking $r$ sufficiently small, one can assume that $U = h^{-1}(B^s \times B^u)$ is the open set in the statement of the Grobman–Hartman theorem (Theorem 4.7). In what follows, we always make these choices.

**Proposition 5.1.** *Let $f \colon \mathbb{R}^n \to \mathbb{R}^n$ be a function of class $C^1$ and let $x_0$ be a hyperbolic critical point of the equation $x' = f(x)$. Then*

$$V^s = \left\{ x \in U : \psi_t(x) \in U \text{ for } t > 0 \right\} \tag{5.4}$$

*and*

$$V^u = \left\{ x \in U : \psi_t(x) \in U \text{ for } t < 0 \right\}. \tag{5.5}$$

*Moreover, for each $t > 0$ we have*

$$\psi_t(V^s) \subset V^s \quad \text{and} \quad \psi_{-t}(V^u) \subset V^u. \tag{5.6}$$

**Proof.** We first observe that

$$E^s \supset \left\{ x \in V : \varphi_t(x) \in V \text{ for } t > 0 \right\}$$

and

$$E^u \supset \left\{ x \in V : \varphi_t(x) \in V \text{ for } t < 0 \right\}.$$

This follows from the Jordan canonical form of the matrix $d_{x_0}f$. Hence, by (5.3), we obtain

$$E^s \cap V = \left\{ x \in V : \varphi_t(x) \in V \text{ for } t > 0 \right\}$$

and

$$E^u \cap V = \left\{ x \in V : \varphi_t(x) \in V \text{ for } t < 0 \right\}.$$

Using (5.1) and (5.2) one can write

$$
\begin{aligned}
V^s &= h^{-1}(E^s \cap V) \\
&= \left\{ h^{-1}(x) \in h^{-1}(V) : (\varphi_t \circ h)(h^{-1}(x)) \in V \text{ for } t > 0 \right\} \\
&= \left\{ h^{-1}(x) \in U : (h \circ \psi_t)(h^{-1}(x)) \in V \text{ for } t > 0 \right\} \\
&= \left\{ h^{-1}(x) \in U : \psi_t(h^{-1}(x)) \in U \text{ for } t > 0 \right\},
\end{aligned}
$$

which establishes (5.4). Identity (5.5) can be obtained in an analogous manner. Finally, the inclusions in (5.6) follow easily from (5.3). Indeed, applying $h^{-1}$ to the first inclusion we obtain

$$
(h^{-1} \circ \varphi_t)(E^s \cap V) \subset h^{-1}(E^s \cap V) = V^s
$$

for $t > 0$, but since $h^{-1} \circ \varphi_t = \psi_t \circ h^{-1}$, we conclude that

$$
\begin{aligned}
\psi_t(V^s) &= (\psi_t \circ h^{-1})(E^s \cap V) \\
&= (h^{-1} \circ \varphi_t)(E^s \cap V) \subset V^s
\end{aligned}
$$

for $t > 0$. The second inclusion in (5.6) can be obtained in an analogous manner. $\qquad\square$

## 5.2. The Hadamard–Perron theorem

In fact, if $f$ is of class $C^k$, then the sets $V^s$ and $V^u$ in (5.2) are manifolds of class $C^k$, respectively, tangent to the spaces $E^s$ and $E^u$ at the point $x_0$ (see Figure 5.1). This means that in a neighborhood of $x_0$ the sets $V^s$ and $V^u$ are graphs of functions of class $C^k$. This is the content of the following result.

**Theorem 5.2** (Hadamard–Perron). *Let $f \colon \mathbb{R}^n \to \mathbb{R}^n$ be a function of class $C^k$, for some $k \in \mathbb{N}$, and let $x_0$ be a hyperbolic critical point of the equation $x' = f(x)$. Then there exists a neighborhood $B$ of $x_0$ such that the sets $V^s \cap B$ and $V^u \cap B$ are manifolds of class $C^k$ containing $x_0$ and satisfying*

$$
T_{x_0}(V^s \cap B) = E^s \quad \text{and} \quad T_{x_0}(V^u \cap B) = E^u.
$$

We shall prove the theorem only for $k = 1$ (for the general case see, for example, [**2**]). The proof (for $k = 1$) consists of showing that in some neighborhood $B$ of $x_0$ the sets $V^s \cap B$ and $V^u \cap B$ are graphs of functions

$$
\varphi \colon E^s \cap B \to E^u \quad \text{and} \quad \psi \colon E^u \cap B \to E^s \tag{5.7}
$$

of class $C^1$, that is,

$$
\begin{aligned}
V^s \cap B &= \left\{ (x, \varphi(x)) : x \in E^s \cap B \right\}, \\
V^u \cap B &= \left\{ (\psi(x), x) : x \in E^u \cap B \right\}.
\end{aligned}
\tag{5.8}
$$

The argument consists essentially of two parts: first, we show that there exist Lipschitz functions $\varphi$ and $\psi$ (see Definition 1.31) satisfying (5.8), and

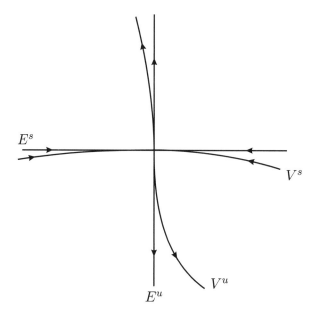

**Figure 5.1.** Manifolds $V^s$ and $V^u$.

then we show that these functions are of class $C^1$. The techniques used in both parts of the proof are also of different nature. For clarity of the presentation, in Section 5.3 we prove that there exist Lipschitz functions $\varphi$ and $\psi$, and we establish their regularity in Section 5.4.

Motivated by Theorem 5.2 we introduce the following notions.

**Definition 5.3.** For a function $f\colon \mathbb{R}^n \to \mathbb{R}^n$ of class $C^1$ and a hyperbolic critical point $x_0$ of the equation $x' = f(x)$, the sets $V^s \cap B$ and $V^u \cap B$ in Theorem 5.2 are called, respectively, *(local) stable manifold* and *(local) unstable manifold* of $x_0$.

Due to the inclusions in (5.6), it is also common to say that the stable and unstable manifolds are *invariant manifolds*.

## 5.3. Existence of Lipschitz invariant manifolds

In this section we establish the existence of Lipschitz functions $\varphi$ and $\psi$ as in (5.7) that satisfy the identities in (5.8).

**Theorem 5.4.** *Let $f\colon \mathbb{R}^n \to \mathbb{R}^n$ be a function of class $C^1$ and let $x_0$ be a hyperbolic critical point of the equation $x' = f(x)$. Then there exist a neighborhood $B$ of $x_0$ and Lipschitz functions*

$$\varphi\colon E^s \cap B \to E^u \quad and \quad \psi\colon E^u \cap B \to E^s \tag{5.9}$$

*satisfying the identities in (5.8).*

**Proof.** We only consider the set $V^s \cap B$, since the result for $V^u \cap B$ can then be obtained in a simple manner. Namely, let $g \colon \mathbb{R}^n \to \mathbb{R}^n$ be the function $g(x) = -f(x)$. We note that if $\psi_t(x_0)$ is the solution of the equation $x' = f(x)$ with $x(0) = x_0$, for $t$ in the maximal interval $I_{x_0} = (a, b)$, then

$$\frac{d}{dt} \psi_{-t}(x_0) = -f(\psi_{-t}(x_0))$$

for every $t \in \mathbb{R}$ and $x_0 \in \mathbb{R}^n$ such that $-t \in I_{x_0}$. This shows that $\bar{\psi}_t(x_0) = \psi_{-t}(x_0)$ is the solution of the equation $x' = g(x)$ with $x(0) = x_0$, for $t \in (-b, -a)$. It follows from the identity $d_{x_0} g = -d_{x_0} f$ that $x_0$ is also a hyperbolic critical point of the equation $x' = g(x)$, with stable and unstable spaces, respectively, $E^u$ and $E^s$. Now let $V_g^s$ and $V_g^u$ be the corresponding sets obtained as in (5.2). By Proposition 5.1, there exists an open set $U \subset \mathbb{R}^n$ such that

$$\begin{aligned} V_g^s &= \big\{ x \in U : \bar{\psi}_t(x) \in U \text{ for } t > 0 \big\} \\ &= \big\{ x \in U : \psi_t(x) \in U \text{ for } t < 0 \big\}, \end{aligned}$$

and hence $V_g^s \cap B = V^u \cap B$ for any sufficiently small neighborhood $B$ of $x_0$. Thus, the result for $V^u \cap B$ follows from the result for $V^s \cap B$ with respect to the function $g$.

Moreover, without loss of generality, from now on we assume that $x_0 = 0$. This can always be obtained through a change of variables (see the proof of Theorem 4.7).

*Step 1. Equivalent formulation of the problem.* Let us consider coordinates $(x, y) \in E^s \times E^u$ (recall that $E^s \oplus E^u = \mathbb{R}^n$). For the equation $x' = f(x)$, it follows from (5.6) that if $(z, \varphi(z)) \in V^s$ is the initial condition at time $t_0 = 0$, then the corresponding solution remains in $V^s$ for all $t > 0$, and thus it is of the form $(x(t), \varphi(x(t)))$, for some function $x(t)$.

As in the proof of Theorem 4.7, we write

$$f(x) = Ax + g(x),$$

where

$$A = d_0 f \quad \text{and} \quad g(x) = f(x) - Ax.$$

We also consider the function $\bar{g}$ in (4.29), which satisfies (4.30). For any sufficiently small $z$, it follows from the Variation of parameters formula in Theorem 2.26 that

$$x(t) = P_s e^{At} z + \int_0^t P_s e^{A(t-\tau)} \bar{g}(x(\tau), \varphi(x(\tau))) \, d\tau \tag{5.10}$$

and

$$\varphi(x(t)) = P_u e^{At} \varphi(z) + \int_0^t P_u e^{A(t-\tau)} \bar{g}(x(\tau), \varphi(x(\tau))) \, d\tau \tag{5.11}$$

for $t \geq 0$ (because then the solution belongs to the neighborhood of zero where $g$ and $\bar{g}$ coincide). The idea of the proof is to use these identities and a fixed point problem to establish the existence of a Lipschitz function $\varphi$.

*Step 2. Behavior of the stable component.* Consider the set $Y$ of all continuous functions $\varphi \colon E^s \to E^u$ such that $\varphi(0) = 0$ and

$$\|\varphi(x) - \varphi(y)\| \leq \|x - y\|, \quad x, y \in E^s. \tag{5.12}$$

It is a complete metric space with the distance

$$d(\varphi, \psi) = \sup\left\{\frac{\|\varphi(x) - \psi(x)\|}{\|x\|} : x \in E^s \setminus \{0\}\right\}$$

(see Exercise 1.14). Given $\varphi \in Y$, let $x = x_\varphi$ be the unique function of class $C^1$ satisfying (5.10). In order to verify that it exists and is unique, we note that since the right-hand side of (5.10) is differentiable in $t$, one can take derivatives to obtain

$$
\begin{aligned}
x'(t) &= AP_s e^{At} z + A \int_0^t P_s e^{A(t-\tau)} \bar{g}(x(\tau), \varphi(x(\tau))) \, d\tau + P_s \bar{g}(x(t), \varphi(x(t))) \\
&= Ax(t) + P_s \bar{g}(x(t), \varphi(x(t))).
\end{aligned}
$$

Now we define

$$h_\varphi(x) = P_s \bar{g}(x, \varphi(x)). \tag{5.13}$$

Taking $\tau = 0$ in (4.12) we obtain $\|P_s\| \leq c$. Hence, it follows from (4.30) and (5.12) that

$$
\begin{aligned}
\|h_\varphi(x) - h_\varphi(y)\| &\leq c\delta\|(x, \varphi(x)) - (y, \varphi(y))\| \\
&= c\delta\|(x - y, \varphi(x) - \varphi(y))\| \\
&\leq c\delta\|x - y\| + c\delta\|\varphi(x) - \varphi(y)\| \\
&\leq 2c\delta\|x - y\|.
\end{aligned}
\tag{5.14}
$$

The Picard–Lindelöf theorem (Theorem 1.18) ensures that there exists a unique solution of the equation

$$x' = Ax + h_\varphi(x) \tag{5.15}$$

for each initial condition with values in $E^s$. By (5.10), the function $x = x_\varphi$ takes only values in $E^s$. Moreover, it follows from (4.12), (4.30) and (5.10) that

$$\|x(t)\| \leq ce^{-\mu t}\|z\| + \int_0^t ce^{-\mu(t-\tau)}\delta\|(x(\tau), \varphi(x(\tau)))\| \, d\tau$$

in the maximal interval of the solution $x(t)$. Since

$$\|\varphi(x(\tau))\| = \|\varphi(x(\tau)) - \varphi(0)\| \leq \|x(\tau)\|,$$

we obtain

$$\|x(t)\| \leq ce^{-\mu t}\|z\| + \int_0^t ce^{-\mu(t-\tau)}2\delta\|x(\tau)\| \, d\tau.$$

Hence, it follows from Gronwall's lemma (Proposition 1.39) that

$$e^{\mu t}\|x(t)\| \le c\|z\|e^{2c\delta t},$$

or equivalently,

$$\|x(t)\| \le ce^{(-\mu+2c\delta)t}\|z\|, \tag{5.16}$$

for $t \ge 0$ in the maximal interval of the solution. Eventually taking $\delta$ smaller so that $-\mu + 2c\delta < 0$, one can proceed as in the proof of Theorem 3.12 to show that the function $x(t)$ is well defined and satisfies (5.16) for every $t \ge 0$.

*Step 3. Some properties of the function $x_\varphi$.* Given $z, \bar{z} \in E^s$, let $x_\varphi$ and $\bar{x}_\varphi$ be the solutions of equation (5.15), respectively, with $x_\varphi(0) = z$ and $\bar{x}_\varphi(0) = \bar{z}$. Since

$$\|\bar{g}(x_\varphi(\tau), \varphi(x_\varphi(\tau))) - \bar{g}(\bar{x}_\varphi(\tau), \varphi(\bar{x}_\varphi(\tau)))\|$$
$$\le \delta\big(\|x_\varphi(\tau) - \bar{x}_\varphi(\tau)\| + \|\varphi(x_\varphi(\tau)) - \varphi(\bar{x}_\varphi(\tau))\|\big) \tag{5.17}$$
$$\le 2\delta\|x_\varphi(\tau) - \bar{x}_\varphi(\tau)\|,$$

writing

$$\rho(t) = \|x_\varphi(t) - \bar{x}_\varphi(t)\|$$

it follows from (5.10) and (4.12) that

$$\rho(t) \le ce^{-\mu t}\|z - \bar{z}\| + \int_0^t ce^{-\mu(t-\tau)}2\delta\rho(\tau)\,d\tau$$

for $t \ge 0$. By Gronwall's lemma, we obtain

$$\rho(t) \le ce^{(-\mu+2c\delta)t}\|z - \bar{z}\| \tag{5.18}$$

for $t \ge 0$.

Given $\varphi, \psi \in Y$, let $x_\varphi$ and $x_\psi$ be, respectively, the solutions of the equations (5.15) and

$$x' = Ax + h_\psi(x)$$

with $x_\varphi(0) = x_\psi(0) = z$. We have

$$\|\bar{g}(x_\varphi(\tau), \varphi(x_\varphi(\tau))) - \bar{g}(x_\psi(\tau), \psi(x_\psi(\tau)))\|$$
$$\le \|\bar{g}(x_\varphi(\tau), \varphi(x_\varphi(\tau))) - \bar{g}(x_\varphi(\tau), \psi(x_\psi(\tau)))\|$$
$$\quad + \|\bar{g}(x_\varphi(\tau), \psi(x_\psi(\tau))) - \bar{g}(x_\psi(\tau), \psi(x_\psi(\tau)))\|$$
$$\le \delta\|\varphi(x_\varphi(\tau)) - \psi(x_\psi(\tau))\| + \delta\|x_\varphi(\tau) - x_\psi(\tau)\| \tag{5.19}$$
$$\le \delta\|\varphi(x_\varphi(\tau)) - \psi(x_\varphi(\tau))\| + \delta\|\psi(x_\varphi(\tau)) - \psi(x_\psi(\tau))\|$$
$$\quad + \delta\|x_\varphi(\tau) - x_\psi(\tau)\|$$
$$\le \delta d(\varphi, \psi)\|x_\varphi(\tau)\| + 2\delta\|x_\varphi(\tau) - x_\psi(\tau)\|.$$

Writing

$$\bar{\rho}(t) = \|x_\varphi(t) - x_\psi(t)\|,$$

it follows from (5.10) and (5.16) that

$$\bar{\rho}(t) \le \int_0^t ce^{-\mu(t-\tau)}\delta d(\varphi,\psi)ce^{(-\mu+2c\delta)\tau}\|z\|\,d\tau + \int_0^t ce^{-\mu(t-\tau)}2\delta\bar{\rho}(\tau)\,d\tau$$

$$\le c^2\delta d(\varphi,\psi)e^{-\mu t}\frac{e^{2c\delta\tau}}{2c\delta}\Big|_{\tau=0}^{\tau=t}\|z\| + \int_0^t ce^{(-\mu+2c\delta)(t-\tau)}2\delta\bar{\rho}(\tau)\,d\tau$$

$$\le \frac{c}{2}d(\varphi,\psi)e^{(-\mu+2c\delta)t}\|z\| + \int_0^t 2c\delta e^{(-\mu+2c\delta)(t-\tau)}\bar{\rho}(\tau)\,d\tau,$$

that is,

$$e^{(\mu-2c\delta)t}\bar{\rho}(t) \le \frac{c}{2}d(\varphi,\psi)\|z\| + \int_0^t 2c\delta e^{(\mu-2c\delta)\tau}\bar{\rho}(\tau)\,d\tau.$$

Hence, it follows from Gronwall's lemma that

$$e^{(\mu-2c\delta)t}\bar{\rho}(t) \le \frac{c}{2}d(\varphi,\psi)\|z\|e^{2c\delta t},$$

or equivalently,

$$\bar{\rho}(t) \le \frac{c}{2}d(\varphi,\psi)\|z\|e^{(-\mu+4c\delta)t}, \tag{5.20}$$

for every $t \ge 0$.

*Step 4. Behavior of the unstable component.* Now we show that

$$\varphi(x_\varphi(t)) = P_u e^{At}\varphi(z) + \int_0^t P_u e^{A(t-\tau)}\bar{g}(x_\varphi(\tau),\varphi(x_\varphi(\tau)))\,d\tau \tag{5.21}$$

for every $t \ge 0$ if and only if

$$\varphi(z) = -\int_0^{+\infty} P_u e^{-A\tau}\bar{g}(x_\varphi(\tau),\varphi(x_\varphi(\tau)))\,d\tau \tag{5.22}$$

for every $t \ge 0$. We first show that the integral in (5.22) is well defined. Indeed, it follows from (4.12), (5.12) and (5.16) that

$$\|P_u e^{-At}\bar{g}(x_\varphi(t),\varphi(x_\varphi(t)))\| \le ce^{-\mu t}2\delta\|x_\varphi(t)\|$$

$$\le ce^{-\mu t}2\delta ce^{(-\mu+2c\delta)t}\|z\| \tag{5.23}$$

$$= 2c^2\delta e^{(-2\mu+2c\delta)t}\|z\|.$$

Since $-2\mu + 2c\delta < 0$, we obtain

$$\int_0^{+\infty}\|P_u e^{-A\tau}\bar{g}(x_\varphi(\tau),\varphi(x_\varphi(\tau)))\|\,d\tau \le 2c^2\delta\int_0^{+\infty}e^{(-2\mu+2c\delta)\tau}\,d\tau < +\infty,$$

which shows that the integral in (5.22) is well defined.

Now we assume that identity (5.21) holds for every $t \ge 0$. Then

$$\varphi(z) = P_u e^{-At}\varphi(x_\varphi(t)) - \int_0^t P_u e^{-A\tau}\bar{g}(x_\varphi(\tau),\varphi(x_\varphi(\tau)))\,d\tau. \tag{5.24}$$

Proceeding as in (5.23), we obtain

$$\|P_u e^{-At}\varphi(x_\varphi(t))\| \le ce^{-\mu t}ce^{(-\mu+2c\delta)t}\|z\|$$
$$\le c^2 e^{(-2\mu+2c\delta)t}\|z\| \to 0$$

when $t \to +\infty$. Thus, letting $t \to +\infty$ in (5.24), we conclude that (5.22) holds. On the other hand, if identity (5.22) holds for every $t \ge 0$, then

$$P_u e^{At}\varphi(z) + \int_0^t P_u e^{A(t-\tau)}\bar{g}(x_\varphi(\tau), \varphi(x_\varphi(\tau)))\, d\tau$$
$$= -\int_t^{+\infty} P_u e^{A(t-\tau)}\bar{g}(x_\varphi(\tau), \varphi(x_\varphi(\tau)))\, d\tau.$$

We show that the right-hand side of this identity is equal to $\varphi(x_\varphi(t))$. Denoting by $F_t(z)$ the solution of equation (5.10), which is also the solution of the autonomous equation (5.15) with initial condition $F_0(z) = z$, it follows from Proposition 1.13 that

$$x_\varphi(\tau) = F_\tau(z) = F_{\tau-t}(F_t(z)) = F_{\tau-t}(x_\varphi(t)).$$

We note that identity (5.22) can be written in the form

$$\varphi(z) = -\int_0^{+\infty} P_u e^{-A\tau}\bar{g}(F_\tau(z), \varphi(F_\tau(z)))\, d\tau. \qquad (5.25)$$

Hence, it follows from (5.25) that

$$-\int_t^{+\infty} P_u e^{A(t-\tau)}\bar{g}(x_\varphi(\tau), \varphi(x_\varphi(\tau)))\, d\tau$$
$$= -\int_t^{+\infty} P_u e^{A(t-\tau)}\bar{g}(F_{\tau-t}(x_\varphi(t)), \varphi(F_{\tau-t}(x_\varphi(t))))\, d\tau$$
$$= -\int_0^{+\infty} P_u e^{-Ar}\bar{g}(F_r(x_\varphi(t)), \varphi(F_r(x_\varphi(t))))\, dr$$
$$= \varphi(x_\varphi(t)).$$

*Step 5. Existence of the function $\varphi$.* Now we use identity (5.22) to show that there exists a unique function $\varphi \in Y$ satisfying (5.21). To that effect, we define a transformation $T$ in $Y$ by

$$T(\varphi)(z) = -\int_0^{+\infty} P_u e^{-A\tau}\bar{g}(x_\varphi(\tau), \varphi(x_\varphi(\tau)))\, d\tau. \qquad (5.26)$$

Since $\bar{g}(0) = 0$ and $x_\varphi(\tau) = 0$ when $z = x_\varphi(0) = 0$, we have $T(\varphi)(0) = 0$. Given $z, \bar{z} \in E^s$, let $x_\varphi$ and $\bar{x}_\varphi$ be again the solutions of equation (5.15), respectively, with $x_\varphi(0) = z$ and $\bar{x}_\varphi(0) = \bar{z}$. It follows from (5.17) and (5.18) that

$$\|\bar{g}(x_\varphi(\tau), \varphi(x_\varphi(\tau))) - \bar{g}(\bar{x}_\varphi(\tau), \varphi(\bar{x}_\varphi(\tau)))\| \le 2c\delta e^{(-\mu+2c\delta)\tau}\|z - \bar{z}\|,$$

and thus,

$$\|T(\varphi)(z) - T(\varphi)(\bar{z})\| \leq \int_0^{+\infty} ce^{-\mu\tau} 2c\delta e^{(-\mu+2c\delta)\tau} \|z - \bar{z}\| \, d\tau$$

$$\leq \frac{c^2\delta}{\mu - c\delta} \|z - \bar{z}\| \leq \|z - \bar{z}\|,$$

for any sufficiently small $\delta$. This guarantees that $T(Y) \subset Y$.

Now we show that $T$ is a contraction. Given $\varphi, \psi \in Y$ and $z \in E^s$, let $x_\varphi$ and $x_\psi$ be the solutions, respectively, of the equations (5.15) and $x' = Ax + h_\psi(x)$ with $x_\varphi(0) = x_\psi(0) = z$. It follows from (5.19), (5.20) and (5.16) that

$$\|\bar{g}(x_\varphi(\tau), \varphi(x_\varphi(\tau))) - \bar{g}(x_\psi(\tau), \psi(x_\psi(\tau)))\|$$

$$\leq \delta d(\varphi, \psi) \|x_\varphi(\tau)\| + c\delta d(\varphi, \psi) \|z\| e^{(-\mu+4c\delta)\tau}$$

$$\leq 2c\delta d(\varphi, \psi) \|z\| e^{(-\mu+4c\delta)\tau},$$

and thus,

$$\|T(\varphi)(z) - T(\psi)(z)\| \leq \int_0^{+\infty} ce^{-\mu\tau} 2c\delta d(\varphi, \psi) \|z\| e^{(-\mu+4c\delta)\tau} \, d\tau$$

$$= \int_0^{+\infty} 2c^2\delta d(\varphi, \psi) \|z\| e^{(-2\mu+4c\delta)\tau} \, d\tau$$

$$\leq \frac{c^2\delta}{\mu - 2c\delta} d(\varphi, \psi) \|z\|.$$

This implies that

$$d(T(\varphi), T(\psi)) \leq \frac{c^2\delta}{\mu - 2c\delta} d(\varphi, \psi). \tag{5.27}$$

For any sufficiently small $\delta$, the transformation $T$ is a contraction. Thus, by Theorem 1.35 there exists a unique function $\varphi \in Y$ such that $T(\varphi) = \varphi$, that is, such that identity (5.22) holds. This shows that there exists a unique function $\varphi \in Y$ satisfying (5.10) and (5.11). Since $\bar{g} = g$ in the ball $B = B(0, r/3)$ (see the proof of Theorem 4.7), we obtain

$$V^s \cap B = \{(z, \varphi(z)) : z \in E^s \cap B\}.$$

This completes the proof of the theorem. $\qquad\square$

It also follows from the proof of Theorem 5.4 that along $V^s \cap B$ the solutions decrease exponentially. More precisely, given $z \in E^s \cap B$, let

$$t \mapsto (x_\varphi(t), \varphi(x_\varphi(t)))$$

be the solution of the equation $x' = f(x)$ with $x_\varphi(0) = z$. It follows from (5.12) and (5.16) that

$$\|(x_\varphi(t), \varphi(x_\varphi(t)))\| \le 2\|x_\varphi(t)\| \le 2ce^{(-\mu+2c\delta)t}\|z\|, \quad t \ge 0.$$

## 5.4. Regularity of the invariant manifolds

In this section we conclude the proof of the Hadamard–Perron theorem (Theorem 5.2). Namely, we show that the functions $\varphi$ and $\psi$ in (5.9) are of class $C^1$ in a neighborhood of $x_0$.

**Theorem 5.5.** *Let $f\colon \mathbb{R}^n \to \mathbb{R}^n$ be a function of class $C^1$ and let $x_0$ be a hyperbolic critical point of the equation $x' = f(x)$. Then the functions $\varphi$ and $\psi$ in Theorem 5.4 are of class $C^1$ in a neighborhood of $x_0$.*

**Proof.** As in the proof of Theorem 5.4, we only consider $V^s \cap B$, since the argument for $V^u \cap B$ is entirely analogous. Moreover, without loss of generality, we assume that $x_0 = 0$.

*Step 1. Construction of an auxiliary transformation.* Let $L(E^s, E^u)$ be the set of all linear transformations from $E^s$ to $E^u$. Also, let $Z$ be the set of all continuous functions $\Phi\colon E^s \to L(E^s, E^u)$ such that $\Phi(0) = 0$ and

$$\sup\left\{\|\Phi(z)\| : z \in E^s\right\} \le 1.$$

Using Proposition 1.30, one can easily verify that equipped with the distance

$$d(\Phi, \Psi) = \sup\left\{\|\Phi(z) - \Psi(z)\| : z \in E^s\right\}$$

the set $Z$ is a complete metric space. Given $\varphi \in Y$ and $z \in E^s$ (see the proof of Theorem 5.4 for the definition of the space $Y$), let $x = x_\varphi$ be the solution of equation (5.15) with $x_\varphi(0) = z$. For simplicity of the exposition, we introduce the notations

$$y_\varphi(\tau, z) = \big(x_\varphi(\tau), \varphi(x_\varphi(\tau))\big)$$

and

$$G(\tau, z, \varphi, \Phi) = \frac{\partial \bar{g}}{\partial x}(y_\varphi(\tau, z)) + \frac{\partial \bar{g}}{\partial y}(y_\varphi(\tau, z))\Phi(x_\varphi(\tau)), \tag{5.28}$$

where $(x, y) \in E^s \times E^u$. Now we define a linear transformation $A(\varphi, \Phi)(z)$ for each $(\varphi, \Phi) \in Y \times Z$ and $z \in E^s$ by

$$A(\varphi, \Phi)(z) = -\int_0^{+\infty} P_u e^{-A\tau} G(\tau, z, \varphi, \Phi) W(\tau)\, d\tau, \tag{5.29}$$

where

$$W = W_{z,\varphi,\Phi}\colon \mathbb{R}_0^+ \to L(E^s, E^s)$$

is the matrix function of class $C^1$ determined by the identity

$$W(t) = P_s e^{At} + \int_0^t P_s e^{A(t-\tau)} G(\tau, z, \varphi, \Phi) W(\tau)\, d\tau. \tag{5.30}$$

Writing

$$C(t, z) = A + P_s G(\tau, z, \varphi, \Phi), \tag{5.31}$$

for each $x \in E^s$ the function $u(t) = W(t)x$ is the solution of the linear equation $u' = C(t, z)u$ with initial condition $u(0) = x$. Since the function $h_\varphi$ in (5.13) is Lipschitz (see (5.14)), it follows from Theorem 1.38 that the maps $(t, z) \mapsto x_\varphi(t)$ and $y_\varphi$ are continuous. Thus,

$$(t, z) \mapsto G(t, z, \varphi, \Phi) \quad \text{and} \quad (t, z) \mapsto C(t, z)$$

are also continuous, since they are compositions of continuous functions (see (5.28)). Hence, the solution $u$ has maximal interval $\mathbb{R}$ and the function $W$ is well defined.

Now we show that the transformation $A$ is well defined. It follows from (4.12) and (4.30) that

$$
\begin{aligned}
B &:= \int_0^{+\infty} \left\| P_u e^{-A\tau} G(\tau, z, \varphi, \Phi) W(\tau) \right\| d\tau \\
&\leq 2c\delta \int_0^{+\infty} e^{-\mu\tau} \|W(\tau)\| d\tau.
\end{aligned}
\tag{5.32}
$$

On the other hand, by (5.30), we have

$$\|W(t)\| \leq ce^{-\mu t} + 2c\delta \int_0^t e^{-\mu(t-\tau)} \|W(\tau)\| d\tau.$$

Multiplying by $e^{\mu t}$, it follows from Gronwall's lemma (Proposition 1.39) that

$$\|W(t)\| \leq ce^{-(\mu - 2c\delta)t} \tag{5.33}$$

for $t \geq 0$, and by (5.32) we conclude that

$$B \leq 2c^2\delta \int_0^{+\infty} e^{(-2\mu + 2c\delta)\tau} d\tau \leq \frac{c^2\delta}{\mu - c\delta}. \tag{5.34}$$

This shows that the transformation $A$ is well defined and

$$\|A(\varphi, \Phi)(z)\| \leq \frac{c^2\delta}{\mu - c\delta} \leq 1$$

for any sufficiently small $\delta$. Since $d_0\bar{g} = d_0 g = 0$, $\varphi(0) = 0$ and $x_\varphi = 0$ for $z = 0$, it follows from (5.29) that $A(\varphi, \Phi)(0) = 0$.

We also show that the function $z \mapsto A(\varphi, \Phi)(z)$ is continuous for each $(\varphi, \Psi) \in Y \times Z$. Given $z, \bar{z} \in E^s$, let $x_\varphi$ and $\bar{x}_\varphi$ be the solutions of equation (5.15), respectively, with $x_\varphi(0) = z$ and $\bar{x}_\varphi(0) = \bar{z}$. We also consider the matrix functions

$$W_z = W_{z,\varphi,\Phi} \quad \text{and} \quad W_{\bar{z}} = W_{\bar{z},\varphi,\Phi}.$$

Given $\varepsilon > 0$, it follows from (5.32) and (5.34) that there exists $T > 0$ such that

$$\int_T^{+\infty} \left\| P_u e^{-A\tau} G(\tau, z, \varphi, \Phi) W_z(\tau) \right\| d\tau < \varepsilon$$

and

$$\int_T^{+\infty} \left\| P_u e^{-A\tau} G(\tau, \bar{z}, \varphi, \Phi) W_{\bar{z}}(\tau) \right\| d\tau < \varepsilon.$$

Thus,

$$\begin{aligned} \|A(\varphi, \Phi)(z) - A(\varphi, \Phi)(\bar{z})\| \\ \leq c \int_0^T e^{-\mu\tau} \|G(\tau, z, \varphi, \Phi) W_z(\tau) - G(\tau, \bar{z}, \varphi, \Phi) W_{\bar{z}}(\tau)\| \, d\tau + 2\varepsilon. \end{aligned} \tag{5.35}$$

Now we show that the function

$$(\tau, z) \mapsto G(\tau, z, \varphi, \Phi) W_z(\tau) \tag{5.36}$$

is continuous. In view of the former discussion, it remains to show that the function $z \mapsto W_z(\tau)$ is continuous. To that effect, we consider the linear equation $u' = C(t, z)u$, with the matrix $C(t, z)$ as in (5.31). Since

$$W_z(t) = \mathrm{Id}_{E^s} + \int_0^t C(s, z) W_z(s) \, ds$$

and

$$W_{\bar{z}}(t) = \mathrm{Id}_{E^s} + \int_0^t C(s, \bar{z}) W_{\bar{z}}(s) \, ds,$$

we have

$$\begin{aligned} W_z(t) - W_{\bar{z}}(t) = \int_0^t [C(s, z) - C(s, \bar{z})] W_z(s) \, ds \\ + \int_0^t C(s, \bar{z})(W_z(s) - W_{\bar{z}}(s)) \, ds. \end{aligned} \tag{5.37}$$

Take $\Delta, r > 0$. Since $(t, z) \mapsto C(t, z)$ is uniformly continuous on the compact set $[0, \Delta] \times \overline{B(0, r)}$, given $\bar{\varepsilon} > 0$, there exists $\bar{\delta} \in (0, r)$ such that

$$\|C(t, z) - C(t, \bar{z})\| < \bar{\varepsilon}$$

for $t \in [0, \Delta]$ and $z, \bar{z} \in \overline{B(0, r)}$ with $\|z - \bar{z}\| < \bar{\delta}$. Moreover,

$$\|C(t, z)\| \leq \lambda, \quad \text{where} \quad \lambda = \|A\| + 2\delta \|P_s\|.$$

Hence, it follows from (5.31), (5.33) and (5.37) that

$$\|W_z(t) - W_{\bar{z}}(t)\| \leq \frac{c\bar{\varepsilon}}{\mu - 2c\delta} + \lambda \int_0^t \|W_z(s) - W_{\bar{z}}(s)\| \, ds$$

for $t \in [0, \Delta]$ and $z, \bar{z} \in \overline{B(0, r)}$ with $\|z - \bar{z}\| < \bar{\delta}$. By Gronwall's lemma, we finally obtain

$$\|W_z(t) - W_{\bar{z}}(t)\| \leq \frac{c\bar{\varepsilon}}{\mu - 2c\delta} e^{\lambda t} \leq \frac{c\bar{\varepsilon}}{\mu - 2c\delta} e^{\lambda \Delta}$$

for $t \in [0, \Delta]$ and $z, \bar{z} \in \overline{B(0, r)}$ with $\|z - \bar{z}\| < \bar{\delta}$. This implies that the function $(t, z) \mapsto W_z(t)$ is continuous in each set $[0, \Delta] \times \overline{B(0, r)}$ and thus also in $\mathbb{R}_0^+ \times E^s$ (we already know that the function is of class $C^1$ in $t$, since it is a solution of the equation $u' = C(t, z)u$). Therefore, the function in (5.36) is also continuous, since it is a composition of continuous functions. Hence, it is uniformly continuous on the compact set $[0, T] \times \overline{B(0, r)}$ and there exists $\delta \in (0, r)$ such that

$$\left\| G(\tau, z, \varphi, \Phi)W_z(\tau) - G(\tau, \bar{z}, \varphi, \Phi)W_{\bar{z}}(\tau) \right\| < \varepsilon$$

for $t \in [0, T]$ and $z, \bar{z} \in \overline{B(0, r)}$ with $\|z - \bar{z}\| < \delta$. It follows from (5.35) that

$$\|A(\varphi, \Phi)(z) - A(\varphi, \Phi)(\bar{z})\| \leq \frac{c\varepsilon}{\mu} + 2\varepsilon$$

for $z, \bar{z} \in \overline{B(0, r)}$ with $\|z - \bar{z}\| < \delta$, and the function $z \mapsto A(\varphi, \Phi)(z)$ is continuous (because it is continuous in each ball $B(0, r)$). This shows that $A(Y \times Z) \subset Z$.

*Step 2. Construction of a fiber contraction.* Now we show that the transformation $S \colon Y \times Z \to Y \times Z$ defined by

$$S(\varphi, \Phi) = (T(\varphi), A(\varphi, \Phi))$$

is a fiber contraction. Given $\varphi \in Y$, $\Phi, \Psi \in Z$ and $z \in E^s$, let

$$W_\Phi = W_{z,\varphi,\Phi} \quad \text{and} \quad W_\Psi = W_{z,\varphi,\Psi}$$

be the matrix functions determined by identity (5.30). We have

$$\|A(\varphi, \Phi)(z) - A(\varphi, \Psi)(z)\|$$

$$\leq c \int_0^{+\infty} e^{-\mu\tau} \left\| G(\tau, z, \varphi, \Phi)W_\Phi(\tau) - G(\tau, z, \varphi, \Psi)W_\Psi(\tau) \right\| d\tau$$

$$\leq c \int_0^{+\infty} e^{-\mu\tau} \delta \big( \|W_\Phi(\tau) - W_\Psi(\tau)\|$$

$$+ \|\Phi(x_\varphi(\tau))W_\Phi(\tau) - \Psi(x_\varphi(\tau))W_\Psi(\tau)\| \big) d\tau$$

$$\leq c \int_0^{+\infty} e^{-\mu\tau} \delta \big( \|W_\Phi(\tau) - W_\Psi(\tau)\|$$

$$+ \|\Phi(x_\varphi(\tau))\| \cdot \|W_\Phi(\tau) - W_\Psi(\tau)\|$$

$$+ \|\Phi(x_\varphi(\tau)) - \Psi(x_\varphi(\tau))\| \cdot \|W_\Psi(\tau)\| \big) d\tau$$

$$\leq c \int_0^{+\infty} e^{-\mu\tau} \delta \big( 2\|W_\Phi(\tau) - W_\Psi(\tau)\| + d(\Phi, \Psi)\|W_\Psi(\tau)\| \big) d\tau.$$

$$(5.38)$$

Now we estimate the norm of

$$w(t) = W_\Phi(t) - W_\Psi(t).$$

In an analogous manner to that in (5.38), using (5.33) we obtain

$$\|w(t)\| \le c \int_0^t e^{-\mu(t-\tau)} 2\delta \|w(\tau)\| \, d\tau + c\delta d(\Phi, \Psi) \int_0^t e^{-\mu(t-\tau)} \|W_\Psi(\tau)\| \, d\tau$$

$$\le 2c\delta \int_0^t e^{-\mu(t-\tau)} \|w(\tau)\| \, d\tau + c^2 \delta d(\Phi, \Psi) \int_0^t e^{-\mu(t-\tau)} e^{-(\mu-2c\delta)\tau} \, d\tau$$

$$\le 2c\delta e^{-\mu t} \int_0^t e^{\mu\tau} \|w(\tau)\| \, d\tau + c^2 \delta d(\Phi, \Psi) e^{-\mu t} \int_0^t e^{2c\delta\tau} \, d\tau$$

$$\le 2c\delta e^{-\mu t} \int_0^t e^{\mu\tau} \|w(\tau)\| \, d\tau + \frac{c}{2} d(\Phi, \Psi) e^{(-\mu+2c\delta)t},$$

which yields the inequality

$$e^{\mu t} \|w(t)\| \le 2c\delta \int_0^t e^{\mu\tau} \|w(\tau)\| \, d\tau + \frac{c}{2} d(\Phi, \Psi) e^{2c\delta t}.$$

For the function

$$\alpha(t) = \frac{c}{2} d(\Phi, \Psi) e^{2c\delta t},$$

we have

$$\frac{\alpha'(t) + 2c\delta e^{\mu t} \|w(t)\|}{\alpha(t) + 2c\delta \int_0^t e^{\mu\tau} \|w(\tau)\| \, d\tau} \le \frac{\alpha'(t)}{\alpha(t)} + \frac{2c\delta e^{\mu t} \|w(t)\|}{\alpha(t) + 2c\delta \int_0^t e^{\mu\tau} \|w(\tau)\| \, d\tau}$$

$$\le \frac{\alpha'(t)}{\alpha(t)} + 2c\delta,$$

and integrating on both sides yields the inequality

$$\log\left(\alpha(t) + 2c\delta \int_0^t e^{\mu\tau} \|w(\tau)\| \, d\tau\right) - \log \alpha(0) \le \log \alpha(t) - \log \alpha(0) + 2c\delta t.$$

Taking exponentials we obtain

$$e^{\mu t} \|w(t)\| \le \alpha(t) + 2c\delta \int_0^t e^{\mu\tau} \|w(\tau)\| \, d\tau \le \alpha(t) e^{2c\delta t},$$

or equivalently,

$$\|W_\Phi(t) - W_\Psi(t)\| \le \frac{c}{2} d(\Phi, \Psi) e^{-(\mu-4c\delta)t}. \tag{5.39}$$

By (5.33) and (5.39), it follows from (5.38) that

$$\|A(\varphi,\Phi)(z) - A(\varphi,\Psi)(z)\| \le c^2\delta d(\Phi,\Psi) \int_0^{+\infty} e^{-2(\mu-2c\delta)\tau}\, d\tau$$

$$+ c^2\delta d(\Phi,\Psi) \int_0^{+\infty} e^{-(2\mu-2c\delta)\tau}\, d\tau$$

$$= 2c^2\delta d(\Phi,\Psi) \int_0^{+\infty} e^{-2(\mu-2c\delta)\tau}\, d\tau$$

$$\le \frac{c^2\delta}{\mu - 2c\delta} d(\Phi,\Psi).$$

This shows that for any sufficiently small $\delta$ the transformation $S$ is a fiber contraction.

*Step 3. Continuity of the transformation $S$.* Given $\varphi,\psi \in Y$, $\Phi \in Z$ and $z \in E^s$, let

$$W_\varphi = W_{z,\varphi,\Phi} \quad \text{and} \quad W_\psi = W_{z,\psi,\Phi}$$

be the matrix functions determined by identity (5.30). We have

$$\|A(\varphi,\Phi)(z) - A(\psi,\Phi)(z)\|$$

$$\le c \int_0^{+\infty} e^{-\mu\tau}\|G(\tau,z,\varphi,\Phi)W_\varphi(\tau) - G(\tau,z,\psi,\Phi)W_\psi(\tau)\|\, d\tau$$

$$\le c \int_0^{+\infty} e^{-\mu\tau}\left\|\frac{\partial\bar{g}}{\partial x}(y_\varphi(\tau,z)) - \frac{\partial\bar{g}}{\partial x}(y_\psi(\tau,z))\right\| \cdot \|W_\varphi(\tau)\|\, d\tau$$

$$+ c \int_0^{+\infty} e^{-\mu\tau}\left\|\frac{\partial\bar{g}}{\partial x}(y_\psi(\tau,z))\right\| \cdot \|W_\varphi(\tau) - W_\psi(\tau)\|\, d\tau$$

$$+ c \int_0^{+\infty} e^{-\mu\tau}\left\|\frac{\partial\bar{g}}{\partial y}(y_\varphi(\tau,z)) - \frac{\partial\bar{g}}{\partial y}(y_\psi(\tau,z))\right\| \cdot \|\Phi(x_\varphi(\tau))\| \cdot \|W_\varphi(\tau)\|\, d\tau$$

$$+ c \int_0^{+\infty} e^{-\mu\tau}\left\|\frac{\partial\bar{g}}{\partial y}(y_\psi(\tau,z))\right\| \cdot \|\Phi(x_\varphi(\tau)) - \Phi(x_\psi(\tau))\| \cdot \|W_\varphi(\tau)\|\, d\tau$$

$$+ c \int_0^{+\infty} e^{-\mu\tau}\left\|\frac{\partial\bar{g}}{\partial y}(y_\psi(\tau,z))\right\| \cdot \|\Phi(x_\psi(\tau))\| \cdot \|W_\varphi(\tau) - W_\psi(\tau)\|\, d\tau,$$

and thus,

$$\|A(\varphi,\Phi)(z) - A(\psi,\Phi)(z)\|$$

$$\le c^2 \int_0^{+\infty} e^{-(2\mu-2c\delta)\tau}\left\|\frac{\partial\bar{g}}{\partial x}(y_\varphi(\tau,z)) - \frac{\partial\bar{g}}{\partial x}(y_\psi(\tau,z))\right\|\, d\tau$$

$$+ c^2 \int_0^{+\infty} e^{-(2\mu-2c\delta)\tau}\left\|\frac{\partial\bar{g}}{\partial y}(y_\varphi(\tau,z)) - \frac{\partial\bar{g}}{\partial y}(y_\psi(\tau,z))\right\|\, d\tau$$

$$+ 2c\delta \int_0^{+\infty} e^{-\mu\tau} \|W_\varphi(\tau) - W_\psi(\tau)\| \, d\tau$$

$$+ c^2\delta \int_0^{+\infty} e^{-(2\mu-2c\delta)\tau} \|\Phi(x_\varphi(\tau)) - \Phi(x_\psi(\tau))\| \, d\tau,$$

using also (5.33). Take $\varepsilon > 0$. It follows from (4.30) that there exists $T > 0$ such that

$$c^2 \int_T^{+\infty} e^{-(2\mu-2c\delta)\tau} \left\| \frac{\partial\bar{g}}{\partial x}(y_\varphi(\tau, z)) - \frac{\partial\bar{g}}{\partial x}(y_\psi(\tau, z)) \right\| \, d\tau$$

$$+ c^2 \int_T^{+\infty} e^{-(2\mu-2c\delta)\tau} \left\| \frac{\partial\bar{g}}{\partial y}(y_\varphi(\tau, z)) - \frac{\partial\bar{g}}{\partial y}(y_\psi(\tau, z)) \right\| \, d\tau$$

$$+ 2c\delta \int_T^{+\infty} e^{-\mu\tau} \|W_\varphi(\tau) - W_\psi(\tau)\| \, d\tau$$

$$+ c^2\delta \int_T^{+\infty} e^{-(2\mu-2c\delta)\tau} \|\Phi(x_\varphi(\tau)) - \Phi(x_\psi(\tau))\| \, d\tau$$

$$\leq 10c^2\delta \int_T^{+\infty} e^{-(2\mu-2c\delta)\tau} \, d\tau < \varepsilon$$

for every $z \in E^s$. Hence,

$$\|A(\varphi, \Phi)(z) - A(\psi, \Phi)(z)\|$$

$$\leq c^2 \int_0^T e^{-(2\mu-2c\delta)\tau} \left\| \frac{\partial\bar{g}}{\partial x}(y_\varphi(\tau, z)) - \frac{\partial\bar{g}}{\partial x}(y_\psi(\tau, z)) \right\| \, d\tau$$

$$+ c^2 \int_0^T e^{-(2\mu-2c\delta)\tau} \left\| \frac{\partial\bar{g}}{\partial y}(y_\varphi(\tau, z)) - \frac{\partial\bar{g}}{\partial y}(y_\psi(\tau, z)) \right\| \, d\tau$$

$$+ c \int_0^T e^{-\mu\tau} \left( \left\| \frac{\partial\bar{g}}{\partial x}(y_\psi(\tau, z)) \right\| + \left\| \frac{\partial\bar{g}}{\partial y}(y_\psi(\tau, z)) \right\| \right) \|W_\varphi(\tau) - W_\psi(\tau)\| \, d\tau$$

$$+ c^2 \int_0^T e^{-(2\mu-2c\delta)\tau} \left\| \frac{\partial\bar{g}}{\partial y}(y_\psi(\tau, z)) \right\| \cdot \|\Phi(x_\varphi(\tau)) - \Phi(x_\psi(\tau))\| \, d\tau + \varepsilon.$$

$$(5.40)$$

Now we recall that the function $\bar{g}$ vanishes outside a closed ball $\overline{B(0, r)}$ (see (4.29)). Hence, in (5.40) it is sufficient to take $\tau \in [0, T]$ and $z \in E^s$ such that

$$\|y_\varphi(\tau, z)\|, \ \|y_\psi(\tau, z)\| < r.$$

On the other hand, again because $\bar{g} = 0$ outside $\overline{B(0, r)}$, for any sufficiently large $R > 0$ the solutions of the equation $x' = Ax + h_\varphi(x)$ (see (5.13)) with initial condition $x(0) \in E^s$ such that $\|x(0)\| > R$, satisfy

$$\|(x(t), \varphi(x(t)))\| > r$$

for every $\varphi \in Y$ and $t \in [0, T]$ (because then they are also solutions of the equation $x' = Ax$). Hence,

$$\|y_\varphi(\tau, z)\|, \ \|y_\psi(\tau, z)\| > r$$

for every $\tau \in [0, T]$ and $z \in E^s$ with $\|z\| > R$. This implies that for $r$ as in the proof of Theorem 4.7 all integrals in (5.40) vanish for $\|z\| > R$. Therefore,

$$d(A(\varphi, \Phi), A(\psi, \Phi))$$

$$\leq c^2 \sup_{z \in B(0,R)} \int_0^T e^{-(2\mu - 2c\delta)\tau} \left\| \frac{\partial \bar{g}}{\partial x}(y_\varphi(\tau, z)) - \frac{\partial \bar{g}}{\partial x}(y_\psi(\tau, z)) \right\| d\tau$$

$$+ c^2 \sup_{z \in B(0,R)} \int_0^T e^{-(2\mu - 2c\delta)\tau} \left\| \frac{\partial \bar{g}}{\partial y}(y_\varphi(\tau, z)) - \frac{\partial \bar{g}}{\partial y}(y_\psi(\tau, z)) \right\| d\tau \qquad (5.41)$$

$$+ 2c\delta \sup_{z \in B(0,R)} \int_0^T e^{-\mu\tau} \|W_\varphi(\tau) - W_\psi(\tau)\| \, d\tau$$

$$+ c^2\delta \sup_{z \in B(0,R)} \int_0^T e^{-(2\mu - 2c\delta)\tau} \|\Phi(x_\varphi(\tau)) - \Phi(x_\psi(\tau))\| \, d\tau + \varepsilon.$$

Since the functions $d\bar{g}$ and $\Phi$ are uniformly continuous on the closed ball $\overline{B(0, 2cR)}$, there exists $\eta > 0$ such that

$$\left\| \frac{\partial \bar{g}}{\partial x}(w) - \frac{\partial \bar{g}}{\partial x}(\bar{w}) \right\| + \left\| \frac{\partial \bar{g}}{\partial y}(w) - \frac{\partial \bar{g}}{\partial y}(\bar{w}) \right\| < \varepsilon \qquad (5.42)$$

and

$$\|\Phi(w) - \Phi(\bar{w})\| < \varepsilon \qquad (5.43)$$

for $w, \bar{w} \in \overline{B(0, 2cR)}$ with $\|w - \bar{w}\| < \eta$. Moreover, it follows from (5.12) and (5.20) that

$$\|y_\varphi(\tau, z) - y_\psi(\tau, z)\| \leq 2\|x_\varphi(\tau) - x_\psi(\tau)\|$$
$$\leq cd(\varphi, \psi)\|z\|e^{(-\mu + 4c\delta)\tau} \leq cRd(\varphi, \psi), \qquad (5.44)$$

for every $\tau \geq 0$ and $z \in E^s$, and any sufficiently small $\delta$ such that $\mu - 4c\delta > 0$. Now we estimate $W_\varphi(t) - W_\psi(t)$. It follows from (5.30) that

$$\|W_\varphi(t) - W_\psi(t)\|$$

$$\leq c \int_0^t e^{-\mu(t-\tau)} \|G(\tau, z, \varphi, \Phi)W_\varphi(\tau) - G(\tau, z, \psi, \Phi)W_\psi(\tau)\| \, d\tau$$

$$\leq c^2 \int_0^t e^{-(2\mu-2c\delta)\tau} \left\| \frac{\partial \bar{g}}{\partial x}(y_\varphi(\tau,z)) - \frac{\partial \bar{g}}{\partial x}(y_\psi(\tau,z)) \right\| d\tau$$

$$+ c^2 \int_0^t e^{-(2\mu-2c\delta)\tau} \left\| \frac{\partial \bar{g}}{\partial y}(y_\varphi(\tau,z)) - \frac{\partial \bar{g}}{\partial y}(y_\psi(\tau,z)) \right\| d\tau$$

$$+ 2c\delta \int_0^t e^{-\mu\tau} \| W_\varphi(\tau) - W_\psi(\tau) \| d\tau$$

$$+ c^2\delta \int_0^t e^{-(2\mu-2c\delta)\tau} \| \Phi(x_\varphi(\tau)) - \Phi(x_\psi(\tau)) \| d\tau.$$

By (5.16), we have

$$\| x_\varphi(\tau) \| \leq c\|z\| \leq cR,$$

and thus,

$$\| y_\varphi(\tau,z) \| \leq 2c\|z\| \leq 2cR$$

for $\varphi \in Y$, $\tau \geq 0$ and $z \in \overline{B(0,R)}$. Hence, it follows from (5.42), (5.43) and (5.44) that

$$\| W_\varphi(t) - W_\psi(t) \| \leq c^2\varepsilon \int_0^t e^{-(2\mu-2c\delta)\tau} d\tau + c^2\delta\varepsilon \int_0^t e^{-(2\mu-2c\delta)\tau} d\tau$$

$$+ 2c\delta \int_0^t e^{-\mu\tau} \| W_\varphi(\tau) - W_\psi(\tau) \| d\tau$$

$$\leq \frac{c^2(1+\delta)\varepsilon}{2\mu-2c\delta} + 2c\delta \int_0^t \| W_\varphi(\tau) - W_\psi(\tau) \| d\tau$$

for $t \geq 0$, whenever $d(\varphi,\psi) < \eta/(cR)$. By Gronwall's lemma, we obtain

$$\| W_\varphi(t) - W_\psi(t) \| \leq \frac{c^2(1+\delta)\varepsilon}{2\mu-2c\delta} e^{2c\delta t}$$

for $t \geq 0$, and it follows from (5.41), (5.42) and (5.43) that

$$d(A(\varphi,\Phi), A(\psi,\Phi)) \leq \frac{c^2\varepsilon}{2\mu-2c\delta} + \frac{c^3(1+\delta)\delta\varepsilon}{\mu-c\delta} \int_0^T e^{-(\mu-2c\delta)\tau} d\tau + \frac{c^2\delta\varepsilon}{2\mu-2c\delta}$$

$$= \frac{c^2\varepsilon}{2\mu-2c\delta} \left( 1 + \delta + \frac{c(1+\delta)\delta}{2\mu-4c\delta} \right),$$

whenever $d(\varphi,\psi) < \eta/(cR)$. This shows that the transformation $A$ is continuous, and the fiber contraction $S$ is also continuous (we already know that $T$ is a contraction; see (5.27)).

*Step 4. Smoothness of the function $\varphi$.* We first show that if $\psi \in Y$ is of class $C^1$ and $T$ is the transformation in (5.26), then $T(\psi)$ is of class $C^1$ and

$$d_z T(\psi) = A(\psi, d\psi)(z). \tag{5.45}$$

To that effect, we note that if $\psi$ is of class $C^1$, then the function $h_\psi \colon E^s \to E^s$ defined by

$$h_\psi(x) = P_s \bar{g}(x, \psi(x))$$

(see (5.13)) is also of class $C^1$. Thus, it follows from Theorem 1.42 applied to the equation

$$x' = Ax + h_\psi(x)$$

that the function $u$ defined by $u(t, z) = x_\psi(t)$ is of class $C^1$. Moreover, for $\Phi(z) = d_z\psi$ it follows from (5.30) that $W(t) = \partial u/\partial z$. Indeed, substituting this function $W$ in the right-hand side of (5.30) (with $\varphi$ replaced by $\psi$) we obtain

$$P_s e^{At} + \int_0^t P_s e^{A(t-\tau)} \left( \frac{\partial \bar{g}}{\partial x}(y_\psi(\tau, z)) \frac{\partial u}{\partial z} + \frac{\partial \bar{g}}{\partial y}(y_\psi(\tau, z)) d_{x_\varphi(\tau)}\psi \frac{\partial u}{\partial z} \right) d\tau$$

$$= P_s e^{At} + \int_0^t P_s e^{A(t-\tau)} \frac{\partial}{\partial z} \bar{g}(y_\psi(\tau, z)) \, d\tau$$

$$= \frac{\partial}{\partial z} \left( e^{At} z + \int_0^t P_s e^{A(t-\tau)} \bar{g}(y_\psi(\tau, z)) \, d\tau \right)$$

$$= \frac{\partial x_\psi(\tau)}{\partial z} = \frac{\partial u}{\partial z},$$

using also the Variation of parameters formula in Theorem 2.26. It follows from the uniqueness of solutions of a differential equation that $W(t) = \partial u/\partial z$. Moreover, applying Leibniz's rule in (5.30), we obtain

$$A(\psi, d\psi)(z) = -\int_0^{+\infty} d_z \left[ P_u e^{-A\tau} \bar{g}(y_\psi(\tau, z)) \right] d\tau = d_z T(\psi).$$

Since $z \mapsto A(\psi, d\psi)(z)$ is continuous, we conclude that $T(\psi)$ is of class $C^1$.

Now we consider the pair $(\varphi_0, \Phi_0) = (0, 0) \in Y \times Z$. Clearly, $\Phi_0(z) = d_z\varphi_0$. We define recursively a sequence $(\varphi_n, \Phi_n)$ by

$$(\varphi_{n+1}, \Phi_{n+1}) = S(\varphi_n, \Phi_n) = (T(\varphi_n), A(\varphi_n, \Phi_n)).$$

Assuming that $\varphi_n$ is of class $C^1$ and $\Phi_n(z) = d_z\varphi_n$, it follows from (5.45) that the function $\varphi_{n+1}$ is of class $C^1$ and

$$d_z\varphi_{n+1} = d_z T(\varphi_n) = A(\varphi_n, \Phi_n) = \Phi_{n+1}. \qquad (5.46)$$

Moreover, if $\varphi$ and $\Phi$ are, respectively, the fixed points of $T$ and $\Psi \mapsto A(\varphi, \Psi)$, then, by the Fiber contraction theorem (Theorem 1.38), the sequences $\varphi_n$ and $\Phi_n$ converge uniformly respectively to $\varphi$ and $\Phi$ in each compact subset of $E^s$. It follows from (5.46) and Proposition 1.41 that $\varphi$ is differentiable and $d_z\varphi = \Phi(z)$. Since $\Phi$ is continuous, this shows that $\varphi$ is of class $C^1$. $\qquad \square$

## 5.5. Exercises

**Exercise 5.1.** Sketch the proof of Theorem 5.2 for $k = 2$, that is, show that if the function $f$ is of class $C^2$, then there exists a neighborhood $B$ of $x_0$ such that $V^s \cap B$ is the graph of a function of class $C^2$.

**Exercise 5.2.** Consider a diffeomorphism $f\colon \mathbb{R}^2 \to \mathbb{R}^2$ of class $C^1$ such that $f(0) = 0$ and $d_0 f = \left(\begin{smallmatrix} a & 0 \\ 0 & b \end{smallmatrix}\right)$, with $0 < a < 1 < b$. Given $\delta > 0$, let

$$V = \{x \in B(0, \delta) : f^n(x) \in B(0, \delta) \text{ for } n \in \mathbb{N}\}.$$

Show that there exist $\delta$ and $\lambda$ with $a < \lambda < 1$ such that if $(x, y) \in V$, then

$$\|f^n(x, y)\| \le \lambda^n \|(x, y)\|, \quad n \in \mathbb{N}.$$

Hint: Write $f$ in the form

$$f(x, y) = \big(ax + g(x, y), bx + h(x, y)\big)$$

and look for $V$ as a graph

$$V = \{(x, \varphi(x)) : x \in (-\delta, \delta)\}$$

of some Lipschitz function $\varphi\colon (-\delta, \delta) \to \mathbb{R}$ with $\varphi(0) = 0$.

*Part 3*

# Equations in the Plane

# Chapter 6

# Index Theory

In this chapter we introduce the notion of the index of a closed path with respect to a vector field in the plane. On purpose, we do not always present the most general results, since otherwise we would need techniques that fall outside the scope of the book. In particular, we discuss how the index varies with perturbations of the path and of the vector field—the index does not change provided that the perturbations avoid the zeros of the vector field. As an application of the notion of index, we establish the existence of a critical point in the interior of any periodic orbit in the plane (in the sense of Jordan's curve theorem). For additional topics we refer the reader to [4, 9, 19].

## 6.1. Index for vector fields in the plane

In this section we introduce the notion of the index of a closed path with respect to a vector field in the plane, and we discus how it varies with perturbations of the path and of the vector field.

**6.1.1. Basic notions.** We first introduce the notion of regular path.

**Definition 6.1.** A continuous function $\gamma\colon [0,1] \to \mathbb{R}^2$ is called a *regular path* if there exists a function $\alpha\colon (a,b) \to \mathbb{R}^2$ of class $C^1$ in some interval $(a,b) \supset [0,1]$ such that $\alpha(t) = \gamma(t)$ and $\alpha'(t) \neq 0$ for $t \in [0,1]$. The image $\gamma([0,1])$ of a regular path is called a *curve*.

For simplicity of the exposition, we also introduce the following notion.

**Definition 6.2.** A regular path $\gamma\colon [0,1] \to \mathbb{R}^2$ is called a *closed path* if $\gamma(0) = \gamma(1)$ and the restriction $\gamma|(0,1)$ is injective (see Figure 6.1).

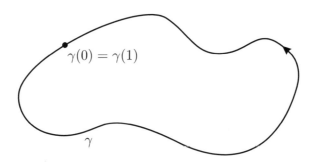

**Figure 6.1.** Closed path.

We note that the image $\gamma([0,1])$ of a closed path has no intersections other than the initial and final points $\gamma(0) = \gamma(1)$.

Consider a vector field $f\colon \mathbb{R}^2 \to \mathbb{R}^2$ of class $C^1$ and write $f = (f_1, f_2)$, where $f_1, f_2\colon \mathbb{R}^2 \to \mathbb{R}$ are the components of $f$. We recall that the *line integral* of $f$ along a regular path $\gamma$ is defined by

$$\int_\gamma f = \int_0^1 f(\gamma(t)) \cdot \gamma'(t)\, dt = \int_0^1 \left[ f_1(\gamma(t))\gamma_1'(t) + f_2(\gamma(t))\gamma_2'(t) \right] dt,$$

where $\gamma(t) = (\gamma_1(t), \gamma_2(t))$.

**Definition 6.3.** Given a closed path $\gamma$ such that $\gamma([0,1])$ contains no zeros of $f$, the number

$$\operatorname{Ind}_f \gamma = \frac{1}{2\pi} \int_\gamma \frac{f_1 \nabla f_2 - f_2 \nabla f_1}{f_1^2 + f_2^2}$$

is called the *index* of $\gamma$ (with respect to $f$).

Now we give a geometric description of the notion of index. To that effect, we note that in a neighborhood of each point $x \in \gamma([0,1])$, the angle between $f(x)$ and the positive part of the horizontal axis is given by

$$\theta(x) = \begin{cases} \arctan(f_2(x)/f_1(x)), & f_1(x) > 0, \\ \pi/2, & f_1(x) = 0 \text{ and } f_2(x) > 0, \\ \arctan(f_2(x)/f_1(x)) + \pi, & f_1(x) < 0, \\ -\pi/2, & f_1(x) = 0 \text{ and } f_2(x) < 0, \end{cases} \qquad (6.1)$$

where arctan denotes the inverse of the tangent with values in $(-\pi/2, \pi/2)$. Now we define

$$\begin{aligned} N(f, \gamma) &= \frac{1}{2\pi} \int_0^1 \frac{d}{dt} \theta(\gamma(t))\, dt \\ &= \frac{1}{2\pi} \int_0^1 \nabla \theta(\gamma(t)) \cdot \gamma'(t)\, dt. \end{aligned} \qquad (6.2)$$

We note that although the function $\theta$ is only defined locally and up to an integer multiple of $2\pi$, the gradient $\nabla\theta$ is defined globally.

**Proposition 6.4.** *If $\gamma\colon [0,1] \to \mathbb{R}^2$ is a closed path such that $\gamma([0,1])$ contains no zeros of $f$, then $\mathrm{Ind}_f\,\gamma = N(f,\gamma)$.*

**Proof.** For the function $\theta$ in (6.1), we have

$$\nabla\theta = \frac{f_1\nabla f_2 - f_2\nabla f_1}{f_1^2 + f_2^2},$$

and thus,

$$
\begin{aligned}
\mathrm{Ind}_f\,\gamma &= \frac{1}{2\pi}\int_\gamma \nabla\theta \\
&= \frac{1}{2\pi}\int_0^1 \nabla\theta(\gamma(t))\cdot\gamma'(t)\,dt \\
&= \frac{1}{2\pi}\int_0^1 \frac{d}{dt}\theta(\gamma(t))\,dt.
\end{aligned}
\tag{6.3}
$$

This yields the desired result. $\qquad\square$

It follows from the definition of $N(f,\gamma)$ in (6.2) and Proposition 6.4 that the index is always an integer.

**Example 6.5.** Consider the phase portraits in Figure 6.2. One can easily verify that

$$\mathrm{Ind}_f\,\gamma_1 = 0, \quad \mathrm{Ind}_f\,\gamma_2 = -1$$

in the first phase portrait, and that

$$\mathrm{Ind}_f\,\gamma_3 = 0, \quad \mathrm{Ind}_f\,\gamma_4 = -1$$

in the second phase portrait.

We note that if a closed path is traversed in the opposite direction, then the index changes sign. More precisely, given a closed path $\gamma\colon [0,1]\to\mathbb{R}^2$, we consider the closed path $-\gamma\colon [0,1]\to\mathbb{R}^2$ defined by

$$(-\gamma)(t) = \gamma(1-t).$$

Since $\gamma$ and $-\gamma$ have the same image, if the curve $\gamma([0,1])$ contains no zeros of $f$, then $\mathrm{Ind}_f(-\gamma)$ is also well defined and

$$\mathrm{Ind}_f(-\gamma) = -\mathrm{Ind}_f\,\gamma.$$

Now we verify that in a sufficiently small neighborhood of a noncritical point the index is always zero.

**Proposition 6.6.** *Let $f\colon \mathbb{R}^2 \to \mathbb{R}^2$ be a function of class $C^1$. If $x_0 \in \mathbb{R}^2$ is such that $f(x_0) \neq 0$, then $\mathrm{Ind}_f\,\gamma = 0$ for any closed path $\gamma$ whose image is contained in a sufficiently small neighborhood of $x_0$.*

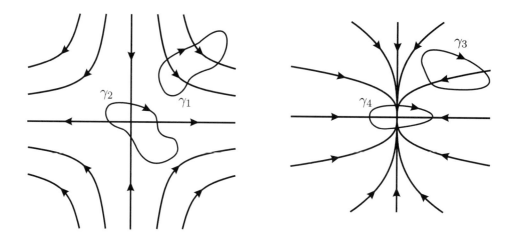

**Figure 6.2.** Phase portraits for Example 6.5.

**Proof.** We first note that by continuity the function $f$ does not take the value zero in a sufficiently small neighborhood of $x_0$. Thus, for any closed path $\gamma$ in this neighborhood, the image $\gamma([0,1])$ contains no zeros of $f$ and the index $\mathrm{Ind}_f\,\gamma$ is well defined. Moreover, also by the continuity of $f$, in a sufficiently small neighborhood of $x_0$ the function $\theta$ in (6.1) only takes values in some interval $[a, b]$ of length less than $2\pi$, and thus, it follows from the first integral in (6.2) that $N(f, \gamma) = 0$. By Proposition 6.4, we conclude that $\mathrm{Ind}_f\,\gamma = 0$. $\qquad\square$

**6.1.2. Perturbations of the path and of the vector field.** In this section we study how the index varies with perturbations of the path $\gamma$ and of the vector field $f$.

We first recall the notion of homotopy.

**Definition 6.7.** Two closed paths $\gamma_0, \gamma_1 \colon [0,1] \to \mathbb{R}^2$ are said to be *homotopic* if there exists a continuous function $H \colon [0,1] \times [0,1] \to \mathbb{R}^2$ such that (see Figure 6.3):

   a) $H(t,0) = \gamma_0(t)$ and $H(t,1) = \gamma_1(t)$ for every $t \in [0,1]$;
   b) $H(0,s) = H(1,s)$ for every $s \in [0,1]$.

We then say that $H$ is a *homotopy* between the paths $\gamma_0$ and $\gamma_1$.

The following result describes how the index varies with homotopies of the path.

**Proposition 6.8.** *The index of a closed path with respect to a vector field $f \colon \mathbb{R}^2 \to \mathbb{R}^2$ of class $C^1$ does not change under homotopies between closed paths whose images contain no zeros of $f$.*

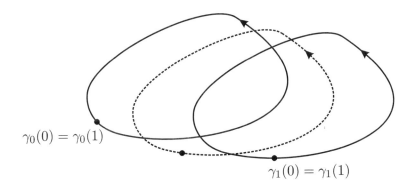

$\gamma_0(0) = \gamma_0(1)$

$\gamma_1(0) = \gamma_1(1)$

**Figure 6.3.** Homotopy between closed paths.

**Proof.** Let $H\colon [0,1] \times [0,1] \to \mathbb{R}^2$ be a homotopy between two closed paths $\gamma_0$ and $\gamma_1$ such that $H([0,1] \times [0,1])$ contains no zeros of $f$. Since the function $(t,s) \mapsto f(\gamma_s(t))$ is uniformly continuous on the compact set $[0,1] \times [0,1]$ (because continuous functions are uniformly continuous on compact sets), given $\varepsilon > 0$ and $s \in [0,1]$, there exists a neighborhood $I_s$ of $s$ in $[0,1]$ such that

$$\|f(\gamma_r(t)) - f(\gamma_s(t))\| < \varepsilon$$

for every $t \in [0,1]$ and $r \in I_s$. This implies that $N(f, \gamma_r) = \operatorname{Ind}_f \gamma_s$ for $r \in I_s$ (because the index is an integer). Hence,

$$\operatorname{Ind}_f \gamma_r = \operatorname{Ind}_f \gamma_s \quad \text{for} \quad r \in I_s.$$

Since the interval $[0,1]$ is compact, there exists a finite subcover $I_{s_1}, \ldots, I_{s_n}$ of $[0,1]$, with $s_1 < s_2 < \cdots < s_n$. Therefore,

$$\operatorname{Ind}_f \gamma_0 = \operatorname{Ind}_f \gamma_{s_1} = \operatorname{Ind}_f \gamma_{s_2} = \cdots = \operatorname{Ind}_f \gamma_{s_n} = \operatorname{Ind}_f \gamma_1,$$

which yields the desired result. $\square$

Now we consider perturbations of the vector field $f$.

**Proposition 6.9.** *Let $F\colon \mathbb{R}^2 \times [0,1] \to \mathbb{R}^2$ be a continuous function such that $x \mapsto F(x,s)$ is of class $C^1$ for each $s \in [0,1]$. If $\gamma$ is a closed path such that $F(x,s) \neq 0$ for every $x \in \gamma([0,1])$ and $s \in [0,1]$, then $\operatorname{Ind}_{F_0} \gamma = \operatorname{Ind}_{F_1} \gamma$, where $F_s(x) = F(x,s)$.*

**Proof.** We proceed in a similar manner to that in the proof of Proposition 6.8. Since the function $(t,s) \mapsto F(\gamma(t),s)$ is uniformly continuous on $[0,1] \times [0,1]$, given $\varepsilon > 0$ and $s \in [0,1]$, there exists a neighborhood $J_s$ of $s$ in $[0,1]$ such that

$$\|F(\gamma(t),r) - F(\gamma(t),s)\| < \varepsilon$$

for every $t \in [0, 1]$ and $r \in J_s$. This implies that $\text{Ind}_{F_r} \gamma = \text{Ind}_{F_s} \gamma$ for $r \in J_s$. Since the interval $[0, 1]$ is compact, there exists a finite subcover $J_{s_1}, \ldots, J_{s_m}$ of $[0, 1]$, with $s_1 < s_2 < \cdots < s_m$. Therefore,

$$\text{Ind}_{F_0} \gamma = \text{Ind}_{F_{s_1}} \gamma = \text{Ind}_{F_{s_2}} \gamma = \cdots = \text{Ind}_{F_{s_m}} \gamma = \text{Ind}_{F_1} \gamma,$$

which yields the desired result.                                                      □

## 6.2. Applications of the notion of index

We give several applications of the notion of index.

### 6.2.1. Periodic orbits and critical points.
In this section we show that any periodic orbit has a critical point in its interior (in the sense of Jordan's curve theorem), as an application of Proposition 6.8. We first recall the notion of a connected set.

**Definition 6.10.** A set $U \subset \mathbb{R}^2$ is said to be *disconnected* if it can be written in the form $U = A \cup B$ for some nonempty sets $A$ and $B$ such that

$$\overline{A} \cap B = A \cap \overline{B} = \varnothing.$$

A set $U \subset \mathbb{R}^2$ is said to be *connected* if it is not disconnected.

We also introduce the notion of a connected component.

**Definition 6.11.** Given $U \subset \mathbb{R}^2$, a set $A \subset U$ is called a *connected component* of $U$ if any connected set $B \subset U$ containing $A$ is equal to $A$.

Now we recall Jordan's curve theorem (for a proof see, for example, [**10**]).

**Proposition 6.12** (Jordan's curve theorem). *If $\gamma\colon [0, 1] \to \mathbb{R}^2$ is a continuous function with $\gamma(0) = \gamma(1)$ such that $\gamma|(0, 1)$ is injective, then $\mathbb{R}^2 \backslash \gamma([0, 1])$ has two connected components, one bounded and one unbounded.*

The bounded connected component in Proposition 6.12 is called the *interior* of the curve $\gamma([0, 1])$. One can now formulate the following result.

**Proposition 6.13.** *Let $f\colon \mathbb{R}^2 \to \mathbb{R}^2$ be a function of class $C^2$ and let $\gamma\colon [0, 1] \to \mathbb{R}^2$ be a closed path. If $\text{Ind}_f \gamma \neq 0$, then the interior of the curve $\gamma([0, 1])$ contains at least one critical point of the equation $x' = f(x)$.*

**Proof.** Let us assume that the interior $U$ of the curve $\gamma([0, 1])$ contains no critical points. Since $\theta$ is of class $C^2$ and $\nabla\theta = (\partial\theta/\partial x, \partial\theta/\partial y)$, it follows from Green's theorem that

$$0 \neq \text{Ind}_f \gamma = \frac{1}{2\pi} \int_\gamma \nabla\theta = \frac{1}{2\pi} \int_{\overline{U}} \left[ \frac{\partial}{\partial x}\left( \frac{\partial\theta}{\partial y} \right) - \frac{\partial}{\partial y}\left( \frac{\partial\theta}{\partial x} \right) \right] dx\, dy = 0.$$

This contradiction shows that $U$ contains at least one critical point.        □

Proposition 6.13 has the following consequence.

**Proposition 6.14.** *If* $f \colon \mathbb{R}^2 \to \mathbb{R}^2$ *is a function of class* $C^2$, *then in the interior of each periodic orbit of the equation* $x' = f(x)$ *there exists at least one critical point.*

**Proof.** Let $\gamma$ be a closed path whose image $\gamma([0,1])$ is a periodic orbit. We first show that $\mathrm{Ind}_f \gamma = \pm 1$. We note that $\mathrm{Ind}_f \gamma = \mathrm{Ind}_g \gamma$, where $g$ is defined in a neighborhood of $\gamma([0,1])$ by $g(x) = f(x)/\|f(x)\|$ (clearly, $f$ does not take the value zero on the periodic orbit). This follows readily from Proposition 6.4, because the function $\theta$ in (6.1) takes the same values for $f$ and $g$. Moreover, without loss of generality, we assume that the curve $\gamma([0,1])$ is contained in the upper half-plane and that the horizontal axis is tangent to $\gamma([0,1])$ at the point $\gamma(0)$, with $\gamma'(0)$ pointing in the positive direction of the axis (if the curve $\gamma$ is traversed in the opposite direction, then one can consider the curve $-\gamma$ and use the identity $\mathrm{Ind}_f \gamma = -\mathrm{Ind}_f(-\gamma)$).

Now we define a function $v \colon \Delta \to \mathbb{R}^2$ in the triangle

$$\Delta = \big\{ (t,s) \in [0,1] \times [0,1] : t \le s \big\}$$

by

$$v(t,s) = \begin{cases} -g(\gamma(0)), & (t,s) = (0,1), \\ g(\gamma(t)), & t = s, \\ (\gamma(s) - \gamma(t))/\|\gamma(s) - \gamma(t)\|, & t < s \text{ and } (t,s) \ne (0,1). \end{cases}$$

One can easily verify that the function $v$ is continuous and that it does not take the value zero. Let $\alpha(t,s)$ be the angle between $v(t,s)$ and the positive part of the horizontal axis. Clearly, $\alpha(0,0) = 0$, because $\gamma'(0)$ is horizontal and points in the positive direction. Moreover, since $\gamma([0,1])$ is contained in the upper half-plane, the function $[0,1] \ni s \mapsto \alpha(0,s)$ varies from 0 to $\pi$. Analogously, the function $[0,1] \ni t \mapsto \alpha(t,1)$ varies from $\pi$ to $2\pi$.

On the other hand, since $v$ does not take the value zero in $\Delta$, it follows from Proposition 6.13 that $\mathrm{Ind}_v \, \partial\Delta = 0$. This shows that the function $[0,1] \ni t \mapsto \alpha(t,t)$ varies from 0 to $2\pi$. Now we observe that $\alpha(t,t)$ coincides with the angle $\theta(\gamma(t))$ between $g(\gamma(t))$ and the positive part of the horizontal axis, by the definition of $v$. Therefore,

$$N(g,\gamma) = \frac{1}{2\pi} \int_0^1 \frac{d}{dt} \theta(\gamma(t)) \, dt = \frac{2\pi}{2\pi} = 1,$$

and it follows from Proposition 6.4 that $\mathrm{Ind}_f \gamma = 1$.

Applying Proposition 6.13, we conclude that there exists at least one critical point in the interior of the periodic orbit. $\qquad\square$

The following is a generalization of Proposition 6.13 to vector fields of class $C^1$ for closed paths whose image is a circle.

**Proposition 6.15.** *Let* $f\colon \mathbb{R}^2 \to \mathbb{R}^2$ *be a function of class* $C^1$ *and let* $\gamma\colon [0,1] \to \mathbb{R}^2$ *be a closed path whose image is a circle. If* $\operatorname{Ind}_f \gamma \neq 0$, *then the interior of the curve* $\gamma([0,1])$ *contains at least one critical point of the equation* $x' = f(x)$.

**Proof.** We assume that the interior of the curve $\gamma([0,1])$ contains no critical points. Now let $\alpha$ be a closed path in the interior of $\gamma([0,1])$ traversed in the same direction as $\gamma$. Then the function $H\colon [0,1] \times [0,1] \to \mathbb{R}^2$ defined by

$$H(s,t) = s\gamma(t) + (1-s)\alpha(t)$$

is a homotopy between $\alpha$ and $\gamma$. Moreover, $H([0,1] \times [0,1])$ contains no critical points. It follows from Proposition 6.8 that

$$\operatorname{Ind}_f \alpha = \operatorname{Ind}_f \gamma \neq 0.$$

On the other hand, if the diameter of the set $\alpha([0,1])$ is sufficiently small, then it follows from Proposition 6.6 that $\operatorname{Ind}_f \alpha = 0$. This contradiction shows that there exists at least one critical point in the interior of the curve $\gamma([0,1])$. $\qquad\square$

**6.2.2. Brouwer's fixed point theorem.** Here and in the following section we provide two applications of Proposition 6.9. The first is a proof of a particular case of Brouwer's fixed point theorem. Namely, we only consider functions of class $C^1$ in the plane.

**Proposition 6.16.** *If* $f\colon \mathbb{R}^2 \to \mathbb{R}^2$ *is a function of class* $C^1$ *and* $B \subset \mathbb{R}^2$ *is a closed ball such that* $f(B) \subset B$, *then* $f$ *has at least one fixed point in* $B$.

**Proof.** Eventually making a change of variables, one can always assume that

$$B = \big\{(x,y) \in \mathbb{R}^2 : x^2 + y^2 \le 1\big\}.$$

Now we consider the transformation $g\colon \mathbb{R}^2 \to \mathbb{R}^2$ defined by $g(x) = x - f(x)$. We want to show that $g$ has zeros in $B$. If there are zeros on the boundary of $B$, then there is nothing to prove. Thus, we assume that there are no zeros on the boundary of $B$, and we consider the function $F\colon \mathbb{R}^2 \times [0,1] \to \mathbb{R}^2$ defined by

$$F(x,t) = tf(x) - x.$$

We note that $F(x,1) = g(x) \neq 0$ for $x \in \partial B$, by hypothesis. Moreover, for $t \in [0,1)$ we have $\|tf(x)\| < 1$, and thus, one cannot have $tf(x) = x$ when $\|x\| = 1$. This shows that $F(x,t) \neq 0$ for every $x \in \partial B$ and $t \in [0,1]$. Now we consider the closed path $\gamma\colon [0,1] \to \mathbb{R}^2$ defined by

$$\gamma(t) = \big(\cos(2\pi t), \sin(2\pi t)\big),$$

that traverses the boundary of $B$ in the positive direction. Then the conditions in Proposition 6.9 are satisfied, and we conclude that

$$\mathrm{Ind}_g \gamma = \mathrm{Ind}_{F_1} \gamma = \mathrm{Ind}_{F_0} \gamma = \mathrm{Ind}_{\mathrm{Id}} \gamma = 1.$$

Hence, it follows from Proposition 6.15 that there exists at least one zero of $g$ in the interior of the ball $B$. $\square$

**6.2.3. Fundamental theorem of algebra.** Now we give a proof of the Fundamental theorem of algebra, again as an application of Proposition 6.9.

**Proposition 6.17** (Fundamental theorem of algebra). *Given* $a_1, \ldots, a_n \in \mathbb{C}$, *the equation* $z^n + a_1 z^{n-1} + \cdots + a_n = 0$ *has at least one root in* $\mathbb{C}$.

**Proof.** Identifying $\mathbb{C}$ with $\mathbb{R}^2$, we define a function $F \colon \mathbb{C} \times [0,1] \to \mathbb{C}$ by

$$F(z,t) = z^n + t(a_1 z^{n-1} + \cdots + a_n).$$

Moreover, given $r > 0$, we consider the closed path $\gamma \colon [0,1] \to \mathbb{C}$ defined by $\gamma(t) = r e^{2\pi i t}$. Now we assume that

$$r > 1 + |a_1| + \cdots + |a_n|.$$

For $z \in \gamma([0,1])$ and $t \in [0,1]$, we have

$$
\begin{aligned}
|z^n| = r^n &> |a_1| r^{n-1} + |a_2| r^{n-1} + \cdots + |a_n| r^{n-1} \\
&> |a_1| r^{n-1} + |a_2| r^{n-2} + \cdots + |a_n| \\
&\geq |a_1 z^{n-1} + \cdots + a_n| \\
&\geq t |a_1 z^{n-1} + \cdots + a_n|,
\end{aligned}
$$

since $r > 1$. This shows that $F(z,t) \neq 0$ for $z \in \gamma([0,1])$ and $t \in [0,1]$. Thus, letting $F_t(z) = F(z,t)$ for each $t \in [0,1]$, it follows from Proposition 6.9 that

$$\mathrm{Ind}_{F_1} \gamma = \mathrm{Ind}_{F_0} \gamma. \tag{6.4}$$

On the other hand, one can easily verify that $\mathrm{Ind}_{F_0} \gamma = n \neq 0$. It follows from (6.4) that $\mathrm{Ind}_{F_1} \gamma \neq 0$. Hence, by Proposition 6.15, there exists at least one zero of $F_1$ in the interior of $\gamma([0,1])$, that is, the polynomial

$$F_1(z) = z^n + a_1 z^{n-1} + \cdots + a_n$$

has at least one root with $|z| < r$. $\square$

## 6.3. Index of an isolated critical point

In this section we introduce the notion of the index of an isolated critical point of the equation $x' = f(x)$, for a vector field $f \colon \mathbb{R}^2 \to \mathbb{R}^2$ of class $C^1$.

**Definition 6.18.** A critical point $x_0$ of the equation $x' = f(x)$ is said to be *isolated* if it is the only critical point in some neighborhood of $x_0$.

Given an isolated critical point $x_0$ of the equation $x' = f(x)$, take $\varepsilon$ sufficiently small such that the ball $B(x_0, \varepsilon)$ contains no critical points besides $x_0$, and consider the closed path $\gamma_\varepsilon \colon [0,1] \to \mathbb{R}^2$ defined by

$$\gamma_\varepsilon(t) = x_0 + \varepsilon\big(\cos(2\pi t), \sin(2\pi t)\big).$$

By Proposition 6.8, the integer number $\operatorname{Ind}_f \gamma_\varepsilon$ is the same for any sufficiently small $\varepsilon > 0$, and one can introduce the following notion.

**Definition 6.19.** Given an isolated critical point $x_0$ of $x' = f(x)$, the *index* of $x_0$ (with respect to $f$) is the integer number $\operatorname{Ind}_f \gamma_\varepsilon$, for any sufficiently small $\varepsilon > 0$. We denote it by $\operatorname{Ind}_f x_0$.

**Example 6.20.** Consider the phase portraits in Figure 2.16. The origin is an isolated critical point in all of them. One can easily verify that the index is $-1$ in the case of the saddle point and $1$ in the remaining phase portraits.

We show that in order to compute the index of any closed path it is sufficient to know the index of the isolated critical points. For simplicity of the proof, we consider only vector fields of class $C^2$.

**Theorem 6.21.** *Let $f \colon \mathbb{R}^2 \to \mathbb{R}^2$ be a function of class $C^2$ and let $\gamma$ be a closed path with positive orientation. If the equation $x' = f(x)$ has finitely many critical points $x_1, \ldots, x_n$ in the interior of the curve $\gamma([0,1])$, then*

$$\operatorname{Ind}_f \gamma = \sum_{i=1}^{n} \operatorname{Ind}_f x_i.$$

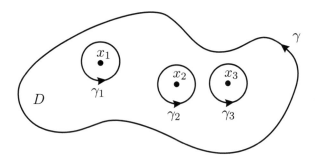

**Figure 6.4.** Paths $\gamma$ and $\gamma_i$ for $i = 1, \ldots, n$.

**Proof.** Let $U$ be the interior of the curve $\gamma([0,1])$ and take $\varepsilon > 0$ sufficiently small such that $\overline{B(x_i, \varepsilon)} \subset U$ for $i = 1, \ldots, n$. We also consider the closed paths $\gamma_i \colon [0,1] \to \mathbb{R}^2$ given by

$$\gamma_i(t) = x_i + \varepsilon\big(\cos(2\pi t), \sin(2\pi t)\big)$$

for $i = 1, \ldots, n$ (see Figure 6.4). It follows from Green's theorem that

$$0 = \int_D \left[ \frac{\partial}{\partial x} \left( \frac{\partial \theta}{\partial y} \right) - \frac{\partial}{\partial y} \left( \frac{\partial \theta}{\partial x} \right) \right] dx\, dy = \int_\gamma \nabla \theta - \sum_{i=1}^n \int_{\gamma_i} \nabla \theta,$$

where $D = \overline{U} \setminus \bigcup_{i=1}^n B(x_i, \varepsilon)$. Thus, by (6.3), we obtain

$$\mathrm{Ind}_f\, \gamma = \frac{1}{2\pi} \int_\gamma \nabla \theta = \sum_{i=1}^n \frac{1}{2\pi} \int_{\gamma_i} \nabla \theta = \sum_{i=1}^n \mathrm{Ind}_f\, \gamma_i.$$

Since the interior of each curve $\gamma_i([0,1])$ contains no critical points besides $x_i$, we have $\mathrm{Ind}_f\, \gamma_i = \mathrm{Ind}_f\, x_i$ for $i = 1, \ldots, n$. Thus,

$$\mathrm{Ind}_f\, \gamma = \sum_{i=1}^n \mathrm{Ind}_f\, \gamma_i = \sum_{i=1}^n \mathrm{Ind}_f\, x_i,$$

which yields the desired result. □

## 6.4. Exercises

**Exercise 6.1.** Consider the equation in polar coordinates

$$\begin{cases} r' = r \cos \theta, \\ \theta' = \sin \theta. \end{cases}$$

a) Sketch the phase portrait.

b) Determine the stability of all critical points.

c) Find the index of all isolated critical points.

**Exercise 6.2.** For each $\alpha, \beta \in \mathbb{R}$, consider the equation

$$\begin{cases} x' = y, \\ y' = -x/4 + \alpha y - \beta(x^2 + 4y^2)y - (x^2 + 4y^2)^2 y. \end{cases}$$

a) Find the index of the origin when $\alpha = 1$.

b) Find whether the equation is conservative when $\alpha \neq 0$.

c) Show that if $\alpha = \beta = 0$, then the origin is stable.

d) Show that if $\alpha > 0$, then there exists at least one periodic orbit.

**Exercise 6.3.** Consider the equation in polar coordinates

$$\begin{cases} r' = r(1 + \cos \theta), \\ \theta' = r(1 - \cos \theta). \end{cases}$$

a) Sketch the phase portrait.

b) Determine the stability of all critical points.

c) Find the index of all isolated critical points.

d) Find whether there are global solutions that are not critical points.

**Exercise 6.4.** Consider the equation

$$\begin{cases} u' = u - uv, \\ v' = uv - v. \end{cases}$$

a) Find the index of the origin.

b) Show that

$$H(u, v) = u + v - \log(uv)$$

is an integral in the quadrant $\{(u, v) \in \mathbb{R}^2 : u, v > 0\}$.

c) Show that there exist infinitely many periodic orbits. Hint: Verify that $H$ has a minimum at $(1, 1)$.

d) Sketch the phase portrait.

**Exercise 6.5.** Show that if $f, g \colon \mathbb{R}^2 \to \mathbb{R}$ are bounded continuous functions, then the equation

$$\begin{cases} x' = y + f(x, y), \\ y' = -x + g(x, y) \end{cases}$$

has at least one critical point.

**Exercise 6.6.** Let $F \colon \mathbb{R} \to \mathbb{R}$ be a bounded continuous function.

a) Show that there exists $x \in \mathbb{R}$ such that $F(-F(x)) = x$. Hint: Use Exercise 6.5.

b) Show that there exists $x \in \mathbb{R}$ such that $\sin(1 - \sin^2(1 - x^2)) = x$.

**Exercise 6.7.** Find the index of the origin for the vector field

$$f(x, y) = \left(2x + y + x^2 + xy^3, x + y - y^2 + x^2y^3\right).$$

**Exercise 6.8.** Let $f \colon \mathbb{R}^2 \to \mathbb{R}^2$ be a function of class $C^1$. Show that if the equation $x' = f(x)$ has a periodic orbit $\gamma$, then the following alternative holds: either $\operatorname{div} f = 0$ in the interior $U$ of $\gamma$ (in the sense of Jordan's curve theorem) or $\operatorname{div} f$ takes different signs in $U$. Hint: Write $f = (f_1, f_2)$ and note that by Green's theorem,

$$\int_U \operatorname{div} f = \int_\gamma (-f_2, f_1).$$

**Exercise 6.9.** Show that if $f, g \colon \mathbb{R} \to \mathbb{R}$ are functions of class $C^1$, then the equation

$$\begin{cases} x' = f(y), \\ y' = g(x) + y^3 \end{cases}$$

has no periodic orbits.

**Exercise 6.10.** Consider the equation

$$x'' = p(x)x' + q(x).$$

Use Exercise 6.8 to show that if $p < 0$, then there are no periodic orbits.

**Exercise 6.11.** Consider the equation

$$\begin{cases} x' = y(1 + x - y^2), \\ y' = x(1 + y - x^2). \end{cases}$$

Show that there are no periodic orbits contained in the first quadrant.

**Exercise 6.12.** Find whether the equation has periodic solutions:

a) $x'' + x^6 + 1 = 0$;

b) $\begin{cases} x' + y^3 - 1 = 0, \\ y' + x^6 + 1 = 0. \end{cases}$

**Solutions.**

 **6.1** b) The origin is the only critical point and is unstable.

   c) 2.

 **6.2** a) 1.

   b) It is not.

 **6.3** b) $(0,0)$ is unstable.

   c) 1.

   d) There are.

 **6.4** a) $-1$.

 **6.7** $-1$.

 **6.12** a) It has not.

   b) It has not.

# Chapter 7

# Poincaré–Bendixson Theory

This chapter is an introduction to the Poincaré–Bendixson theory. After introducing the notion of invariant set, we consider the $\alpha$-limit and $\omega$-limit sets and we establish some of their basic properties. In particular, we show that bounded semiorbits give rise to connected compact $\alpha$-limit and $\omega$-limit sets. We then establish one of the important results of the qualitative theory of differential equations in the plane, the Poincaré–Bendixson theorem, which characterizes the $\alpha$-limit and $\omega$-limit sets of bounded semiorbits. In particular, it allows one to establish a criterion for the existence of periodic orbits. For additional topics we refer the reader to [**9, 13, 15, 17**].

## 7.1. Limit sets

Let the function $f\colon \mathbb{R}^n \to \mathbb{R}^n$ be continuous and locally Lipschitz (see Definition 3.15). Then the equation

$$x' = f(x) \tag{7.1}$$

has unique solutions. We denote by $\varphi_t(x_0)$ the solution with $x(0) = x_0$, for $t \in I_{x_0}$, where $I_{x_0}$ is the corresponding maximal interval.

### 7.1.1. Basic notions. We first introduce the notion of invariant set.

**Definition 7.1.** A set $A \subset \mathbb{R}^n$ is said to be *invariant* (with respect to equation (7.1)) if $\varphi_t(x) \in A$ for every $x \in A$ and $t \in I_x$.

**Example 7.2.** Consider the equation

$$\begin{cases} x' = y, \\ y' = -x. \end{cases} \tag{7.2}$$

Its phase portrait is the one shown in Figure 1.4. The origin and each circle centered at the origin are invariant sets. More generally, any union of circles and any union of circles together with the origin are invariant sets.

We denote the *orbit* of a point $x \in \mathbb{R}^n$ (see Definition 1.50) by

$$\gamma(x) = \gamma_f(x) = \{\varphi_t(x) : t \in I_x\}.$$

It is also convenient to introduce the following notions.

**Definition 7.3.** Given $x \in \mathbb{R}^n$, the set

$$\gamma^+(x) = \gamma_f^+(x) = \{\varphi_t(x) : t \in I_x \cap \mathbb{R}^+\}$$

is called the *positive semiorbit* of $x$, and the set

$$\gamma^-(x) = \gamma_f^-(x) = \{\varphi_t(x) : t \in I_x \cap \mathbb{R}^-\}$$

is called the *negative semiorbit* of $x$.

One can easily verify that a set is invariant if and only if it is a union of orbits. In other words, a set $A \subset \mathbb{R}^n$ is invariant if and only if

$$A = \bigcup_{x \in A} \gamma(x).$$

Now we introduce the notions of $\alpha$-limit and $\omega$-limit sets.

**Definition 7.4.** Given $x \in \mathbb{R}^n$, the *$\alpha$-limit* and *$\omega$-limit sets* of $x$ (with respect to equation (7.1)) are defined respectively by

$$\alpha(x) = \alpha_f(x) = \bigcap_{y \in \gamma(x)} \overline{\gamma^-(y)}$$

and

$$\omega(x) = \omega_f(x) = \bigcap_{y \in \gamma(x)} \overline{\gamma^+(y)}.$$

**Example 7.5** (continuation of Example 7.2). For equation (7.2), we have

$$\alpha(x) = \omega(x) = \gamma^+(x) = \gamma^-(x) = \gamma(x)$$

for any $x \in \mathbb{R}^2$.

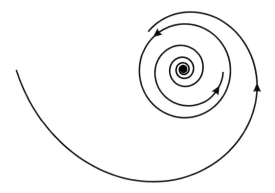

**Figure 7.1.** Phase portrait of equation (7.3).

**Example 7.6.** Consider the equation in polar coordinates

$$\begin{cases} r' = r(1 - r), \\ \theta' = 1. \end{cases} \tag{7.3}$$

Its phase portrait is the one shown in Figure 7.1. Let

$$S = \left\{ x \in \mathbb{R}^2 : \|x\| = 1 \right\}.$$

We have

$$\alpha(x) = \omega(x) = \gamma^+(x) = \gamma^-(x) = \gamma(x)$$

for $x \in \{(0,0)\} \cup S$,

$$\alpha(x) = \{(0,0)\} \quad \text{and} \quad \omega(x) = S$$

for $x \in \mathbb{R}^2$ with $0 < \|x\| < 1$, and finally,

$$\alpha(x) = \varnothing \quad \text{and} \quad \omega(x) = S$$

for $x \in \mathbb{R}^2$ with $\|x\| > 1$.

**Example 7.7.** For the phase portrait in Figure 7.2 we have:

$$\alpha(x_1) = \omega(x_1) = \varnothing;$$

$$\alpha(x_2) = \varnothing, \quad \omega(x_2) = \{q\};$$

$$\alpha(x_3) = \omega(x_3) = \{p\};$$

$$\alpha(x_4) = \omega(x_4) = \gamma(x_4).$$

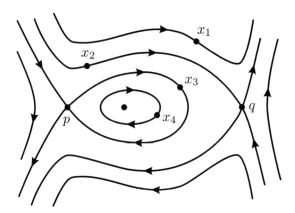

**Figure 7.2.** Phase portrait for Example 7.7.

**7.1.2. Additional properties.** In this section we establish some properties of the $\alpha$-limit and $\omega$-limit sets for equation (7.1).

**Proposition 7.8.** *If the positive semiorbit $\gamma^+(x)$ of a point $x \in \mathbb{R}^n$ is bounded, then:*

a) *$\omega(x)$ is compact, connected and nonempty;*

b) *$y \in \omega(x)$ if and only if there exists a sequence $t_k \nearrow +\infty$ such that $\varphi_{t_k}(x) \to y$ when $k \to \infty$;*

c) *$\varphi_t(y) \in \omega(x)$ for every $y \in \omega(x)$ and $t > 0$;*

d) *$\inf\{\|\varphi_t(x) - y\| : y \in \omega(x)\} \to 0$ when $t \to +\infty$.*

**Proof.** Let $K = \overline{\gamma^+(x)}$. It follows readily from the definition of $\omega$-limit set that $\omega(x)$ is closed. On the other hand, $\omega(x) \subset K$ and thus $\omega(x)$ is also bounded. Hence, the $\omega$-limit set is compact.

Moreover, since the semiorbit $\gamma^+(x)$ is bounded, we have $\mathbb{R}^+ \subset I_x$ (by Theorem 1.46), and thus,

$$\omega(x) = \bigcap_{t>0} A_t, \tag{7.4}$$

where

$$A_t = \overline{\{\varphi_s(x) : s > t\}}.$$

Identity (7.4) yields the second property in the proposition. Indeed, if $y \in \omega(x)$, then there exists a sequence $t_k \nearrow +\infty$ such that $y \in A_{t_k}$ for $k \in \mathbb{N}$. Thus, there is also a sequence $s_k \nearrow +\infty$ with $s_k \geq t_k$ for $k \in \mathbb{N}$ such that $\varphi_{s_k}(x) \to y$ when $k \to \infty$. On the other hand, if there exists a sequence $t_k \nearrow +\infty$ as in the second property in the proposition, then $y \in A_{t_k}$ for

$k \in \mathbb{N}$, and hence,

$$y \in \bigcap_{k=1}^{\infty} A_{t_k} = \bigcap_{t>0} A_t,$$

because $A_t \subset A_{t'}$ for $t > t'$.

Now we consider a sequence $(\varphi_k(x))_k$ contained in the compact set $K$. By compactness, there exists a subsequence $(\varphi_{t_k}(x))_k$, with $t_k \nearrow +\infty$, converging to a point of $K$. This shows that $\omega(x)$ is nonempty.

Now we show that $\omega(x)$ is connected. Otherwise, by Definition 6.10, we would have $\omega(x) = A \cup B$ for some nonempty sets $A$ and $B$ such that $\overline{A} \cap B = A \cap \overline{B} = \varnothing$. Since $\omega(x)$ is closed, we have

$$\overline{A} = \overline{A} \cap \omega(x) = \overline{A} \cap (A \cup B)$$
$$= (\overline{A} \cap A) \cup (\overline{A} \cap B) = A$$

and analogously $\overline{B} = B$. This shows that the sets $A$ and $B$ are closed, and hence, they are at a positive distance, that is,

$$\delta := \inf \left\{ \|a - b\| : a \in A, b \in B \right\} > 0.$$

Now we consider the set

$$C = \left\{ z \in \mathbb{R}^2 : \inf_{y \in \omega(x)} \|z - y\| \geq \frac{\delta}{4} \right\}.$$

One can easily verify that $C \cap K$ is compact and nonempty. Hence, it follows from the second property in the proposition that $C \cap K \cap \omega(x) \neq \varnothing$. But by the definition of the set $C$ we know that $C \cap K$ does not intersect $\omega(x)$. This contradiction shows that $\omega(x)$ is connected.

In order to verify the third property in the proposition, we recall that, by the second property, if $y \in \omega(x)$, then there exists a sequence $t_k \nearrow +\infty$ such that $\varphi_{t_k}(x) \to y$ when $k \to \infty$. By Theorem 1.40, the function $y \mapsto \varphi_t(y)$ is continuous for each fixed $t$. Thus, given $t > 0$, we have

$$\varphi_{t_k+t}(y) = \varphi_t(\varphi_{t_k}(y)) \to \varphi_t(y)$$

when $k \to \infty$. Since $t_k + t \nearrow +\infty$ when $k \to \infty$, it follows from the second property that $\varphi_t(y) \in \omega(x)$.

Finally, we establish the last property in the proposition. Otherwise, there would exist a sequence $t_k \nearrow +\infty$ and a constant $\delta > 0$ such that

$$\inf_{y \in \omega(x)} \|\varphi_{t_k}(x) - y\| \geq \delta \tag{7.5}$$

for $k \in \mathbb{N}$. Since the set $K$ is compact, there exists a convergent subsequence $(\varphi_{t'_k}(x))_k$ of $(\varphi_{t_k}(x))_k \subset K$, which by the second property in the proposition has a limit $p \in \omega(x)$. On the other hand, it follows from (7.5) that

$$\|\varphi_{t'_k}(x) - y\| \geq \delta$$

for every $y \in \omega(x)$ and $k \in \mathbb{N}$. Thus, $\|p - y\| \geq \delta$ for $y \in \omega(x)$, which implies that $p \notin \omega(x)$. This contradiction yields the desired result. $\qquad\square$

We have an analogous result for the $\alpha$-limit set.

**Proposition 7.9.** *If the negative semiorbit $\gamma^-(x)$ of a point $x \in \mathbb{R}^n$ is bounded, then:*

    a) *$\alpha(x)$ is compact, connected and nonempty;*

    b) *$y \in \alpha(x)$ if and only if there exists a sequence $t_k \searrow -\infty$ such that $\varphi_{t_k}(x) \to y$ when $k \to \infty$;*

    c) *$\varphi_t(y) \in \alpha(x)$ for every $y \in \alpha(x)$ and $t < 0$;*

    d) *$\inf\{\|\varphi_t(x) - y\| : y \in \alpha(x)\} \to 0$ when $t \to -\infty$.*

**Proof.** As in the proof of Theorem 5.4, let $g \colon \mathbb{R}^n \to \mathbb{R}^n$ be the function $g(x) = -f(x)$. We recall that if $\varphi_t(x_0)$ is the solution of the equation $x' = f(x)$ with $x(0) = x_0$, for $t$ in the maximal interval $I_{x_0} = (a, b)$, then $\psi_t(x_0) = \varphi_{-t}(x_0)$ is the solution of the equation $x' = g(x)$ with $x(0) = x_0$, for $t \in (-b, -a)$. This implies that

$$\gamma_g(x) = \gamma_f(x) \quad \text{and} \quad \gamma_g^+(x) = \gamma_f^-(x) \tag{7.6}$$

for every $x \in \mathbb{R}^n$, and thus,

$$\alpha_f(x) = \bigcap_{y \in \gamma_f(x)} \overline{\gamma_f^-(y)} = \bigcap_{y \in \gamma_g(x)} \overline{\gamma_g^+(x)} = \omega_g(x). \tag{7.7}$$

Now we assume that the negative semiorbit $\gamma_f^-(x)$ is bounded. Since $\gamma_g^+(x) = \gamma_f^-(x)$, it follows from Proposition 7.8 that:

    a) $\omega_g(x)$ is compact, connected and nonempty;

    b) $y \in \omega_g(x)$ if and only if there exists a sequence $t_k \nearrow +\infty$ such that $\psi_{t_k}(x) \to y$ when $k \to \infty$;

    c) $\psi_t(y) \in \omega_g(x)$ for every $y \in \omega_g(x)$ and $t > 0$;

    d) $\inf\{\|\psi_t(x) - y\| : y \in \omega_g(x)\} \to 0$ when $t \to +\infty$.

Since $\psi_t = \varphi_{-t}$, the proposition now follows readily from (7.7) and these four properties. $\qquad\square$

## 7.2. The Poincaré–Bendixson theorem

Now we turn to $\mathbb{R}^2$ and we establish one of the important results of the qualitative theory of differential equations: the Poincaré–Bendixson theorem.

**7.2.1. Intersections with transversals.** We first establish an auxiliary result. Let $f \colon \mathbb{R}^2 \to \mathbb{R}^2$ be a function of class $C^1$. Also, let $L$ be a line segment *transverse* to $f$. This means that for each $x \in L$ the directions of $L$ and $f(x)$ generate $\mathbb{R}^2$ (see Figure 7.3). We then say that $L$ is a *transversal* to $f$.

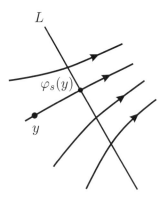

**Figure 7.3.** Orbits in the neighborhood of a transversal.

**Proposition 7.10.** *Given $x \in \mathbb{R}^2$, the intersection $\omega(x) \cap L$ contains at most one point.*

**Proof.** Assume that $\omega(x) \cap L$ is nonempty and take $q \in \omega(x) \cap L$. By Proposition 7.8, there exists a sequence $t_k \nearrow +\infty$ such that $\varphi_{t_k}(x) \to q$ when $k \to \infty$. On the other hand, since $L$ is a transversal to $f$, it follows from the Flow box theorem (Theorem 1.54) that for each $y \in \mathbb{R}^2$ sufficiently close to $L$ there exists a unique time $s$ such that $\varphi_s(y) \in L$ and $\varphi_t(y) \notin L$ for $t \in (0, s)$ when $s > 0$, or for $t \in (s, 0)$ when $s < 0$ (see Figure 7.3); in particular, for each $k \in \mathbb{N}$ there exists $s = s_k$ as above such that $x_k = \varphi_{t_k + s_k}(x) \in L$.

Now we consider two cases: either $(x_k)_k$ is a constant sequence, in which case the orbit of $x$ is periodic, or $(x_k)_k$ is not a constant sequence. In the first case, since the orbit of $x$ is periodic, the $\omega$-limit set $\omega(x) = \gamma(x)$ only intersects $L$ at the constant value of the sequence $(x_k)_k$, and thus $\omega(x) \cap L = \{q\}$. In the second case, let us consider two successive points of intersection $x_k$ and $x_{k+1}$, that can be disposed in $L$ in the two forms in Figure 7.4. We note that along $L$ the vector field $f$ always points to the same side (in other words, the projection of $f$ on the perpendicular to $L$ always has the same direction). Otherwise, since $f$ is continuous, there would exist at least one point $z \in L$ with $f(z)$ equal to zero or with the direction of $L$, but then $L$ would not be a transversal. We also note that the segment of orbit between $x_k$ and $x_{k+1}$ together with the line segment between these two points form a continuous curve $C$ whose complement

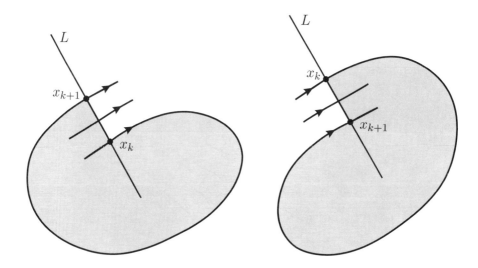

**Figure 7.4.** Intersections with the transversal $L$.

$\mathbb{R}^2 \setminus C$ has two connected components, as a consequence of Jordan's curve theorem (Proposition 6.12). The bounded connected component is marked in gray in Figure 7.4. Due to the direction of $f$ on the segment between $x_k$ and $x_{k+1}$ (see Figure 7.4), the positive semiorbit $\gamma^+(x_k)$ is contained in the unbounded connected component. This implies that the next intersection $x_{k+2}$ does not belong to the line segment between $x_k$ and $x_{k+1}$. Therefore, the points $x_k$, $x_{k+1}$ and $x_{k+2}$ are ordered on the transversal $L$ as shown in Figure 7.5. Due to the monotonicity of the sequence $(x_k)_k$ along $L$, it has at most one accumulation point in $L$ and hence $\omega(x) \cap L = \{q\}$. $\qquad\square$

**7.2.2. The Poincaré–Bendixson theorem.** The following is the main result of this chapter.

**Theorem 7.11** (Poincaré–Bendixson). *Let* $f \colon \mathbb{R}^2 \to \mathbb{R}^2$ *be a function of class* $C^1$. *For equation* (7.1), *if the positive semiorbit* $\gamma^+(x)$ *of a point* $x$ *is bounded and* $\omega(x)$ *contains no critical points, then* $\omega(x)$ *is a periodic orbit.*

**Proof.** Since the semiorbit $\gamma^+(x)$ is bounded, it follows from Proposition 7.8 that $\omega(x)$ is nonempty. Take a point $p \in \omega(x)$. It follows from the first and third properties in Proposition 7.8, together with the definition of $\omega$-limit set, that $\omega(p)$ is nonempty and $\omega(p) \subset \omega(x)$. Now take a point $q \in \omega(p)$. By hypothesis, $q$ is not a critical point, and thus there exists a line segment $L$ containing $q$ that is transverse to $f$. Since $q \in \omega(p)$, by the second property in Proposition 7.8, there exists a sequence $t_k \nearrow +\infty$ such that $\varphi_{t_k}(p) \to q$ when $k \to \infty$. Proceeding as in the proof of Proposition 7.10, one can always

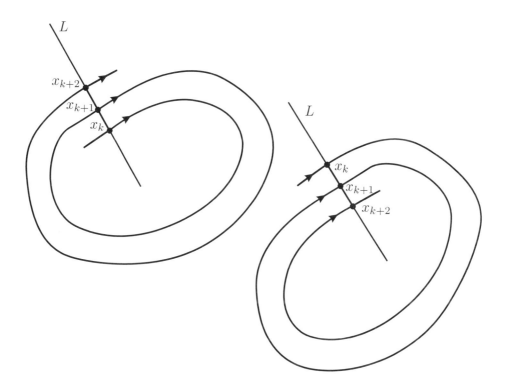

**Figure 7.5.** Intersections $x_k$, $x_{k+1}$ and $x_{k+2}$ with the transversal $L$.

assume that $\varphi_{t_k}(p) \in L$ for $k \in \mathbb{N}$. On the other hand, since $p \in \omega(x)$, it follows from the third property in Proposition 7.8 that $\varphi_{t_k}(p) \in \omega(x)$ for $k \in \mathbb{N}$. Since $\varphi_{t_k}(p) \in \omega(x) \cap L$, by Proposition 7.10 we obtain

$$\varphi_{t_k}(p) = q \quad \text{for every} \quad k \in \mathbb{N}.$$

This implies that $\gamma(p)$ is a periodic orbit. In particular, $\gamma(p) \subset \omega(x)$.

It remains to show that $\omega(x) = \gamma(p)$. If $\omega(x) \setminus \gamma(p) \neq \varnothing$, then since $\omega(x)$ is connected, in each neighborhood of $\gamma(p)$ there exist points of $\omega(x)$ that are not in $\gamma(p)$. We note that any neighborhood of $\gamma(p)$ that is sufficiently small contains no critical points. Thus, there exists a transversal $L'$ to $f$ containing one of these points, which is in $\omega(x)$, and a point of $\gamma(p)$. That is, $\omega(x) \cap L'$ contains at least two points, because $\gamma(p) \subset \omega(x)$; but this contradicts Proposition 7.10. Therefore, $\omega(x) = \gamma(p)$ and the $\omega$-limit set of $x$ is a periodic orbit. $\qquad\square$

One can obtain an analogous result to Theorem 7.11 for bounded negative semiorbits.

**Theorem 7.12.** *Let $f\colon \mathbb{R}^2 \to \mathbb{R}^2$ be a function of class $C^1$. For equation (7.1), if the negative semiorbit $\gamma^-(x)$ of a point $x$ is bounded and $\alpha(x)$ contains no critical points, then $\alpha(x)$ is a periodic orbit.*

**Proof.** As in the proof of Proposition 7.9, consider the function $g\colon \mathbb{R}^2 \to \mathbb{R}^2$ defined by $g(x) = -f(x)$ and the equation $x' = g(x)$. By (7.6) and (7.7), we have

$$\gamma_f^-(x) - \gamma_g^+(x) \quad \text{and} \quad \alpha_f(x) = \omega_g(x).$$

The result is now an immediate consequence of Theorem 7.11. $\qquad\square$

**Example 7.13** (continuation of Example 7.6). We already know that equation (7.3) has a periodic orbit (see Figure 7.1), namely the circle of radius 1 centered at the origin. Now we deduce the existence of a periodic orbit as an application of the Poincaré–Bendixson theorem. Consider the ring

$$D = \left\{ x \in \mathbb{R}^2 : \frac{1}{2} < \|x\| < 2 \right\}.$$

For $r = 1/2$ we have $r' = 1/4 > 0$, and for $r = 2$ we have $r' = 2 < 0$. This implies that any orbit entering $D$ no longer leaves $D$ (for positive times). This corresponds to the qualitative behavior shown in Figure 7.6. In particular, any positive semiorbit $\gamma^+(x)$ of a point $x \in D$ is contained in $D$

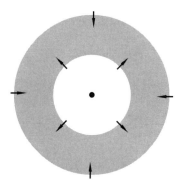

**Figure 7.6.** Behavior in the boundary of $D$.

and hence it is bounded. Moreover, it follows from (7.3) that the origin is the only critical point. By the Poincaré–Bendixson theorem (Theorem 7.11), we conclude that $\omega(x)$ is a periodic orbit for each $x \in D$.

**Example 7.14.** Consider the equation

$$\begin{cases} x' = x(x^2 + y^2 - 3x - 1) - y, \\ y' = y(x^2 + y^2 - 3x - 1) + x, \end{cases} \tag{7.8}$$

which in polar coordinates takes the form

$$\begin{cases} r' = r(r^2 - 3r\cos\theta - 1), \\ \theta' = 1. \end{cases}$$

For any sufficiently small $r$, we have

$$r^2 - 3r\cos\theta - 1 < 0,$$

and thus $r' < 0$. Moreover, for any sufficiently large $r$, we have

$$r^2 - 3r\cos\theta - 1 > 0,$$

and thus $r' > 0$. On the other hand, the origin is the only critical point. Now we use an analogous argument to that in Example 7.13. Namely, for $r_1 > 0$ sufficiently small and $r_2 > 0$ sufficiently large, there are no critical points in the ring

$$D' = \{x \in \mathbb{R}^2 : r_1 < \|x\| < r_2\}.$$

Moreover, any negative semiorbit $\gamma^-(x)$ of a point $x \in D'$ is contained in $D'$, and hence it is bounded. It follows from Theorem 7.12 that $\alpha(x) \subset D'$ is a periodic orbit for each $x \in D'$. In particular, equation (7.8) has at least one periodic orbit in $D'$.

Now we formulate a result generalizing the Poincaré–Bendixson theorem to the case when $\omega(x)$ contains critical points.

**Theorem 7.15.** *Let $f\colon \mathbb{R}^2 \to \mathbb{R}^2$ be a function of class $C^1$. For equation (7.1), if the positive semiorbit $\gamma^+(x)$ of a point $x$ is contained in a compact set where there are at most finitely many critical points, then one of the following alternatives holds:*

a) *$\omega(x)$ is a critical point;*

b) *$\omega(x)$ is a periodic orbit;*

c) *$\omega(x)$ is a union of a finite number of critical points and homoclinic or heteroclinic orbits.*

**Proof.** Since $\omega(x) \subset \overline{\gamma^+(x)}$, the set $\omega(x)$ contains at most finitely many critical points. If it only contains critical points, then it is necessarily a single critical point, since by Proposition 7.8 the set $\omega(x)$ is connected.

Now we assume that $\omega(x)$ contains noncritical points and that it contains at least one periodic orbit $\gamma(p)$. We show that $\omega(x)$ is the periodic orbit. Otherwise, since $\omega(x)$ is connected, there would exist a sequence $(x_k)_k \subset \omega(x) \setminus \gamma(p)$ and a point $x_0 \in \gamma(p)$ such that $x_k \to x_0$ when $k \to \infty$. Now we consider a transversal $L$ to the vector field $f$ such that $x_0 \in L$. It follows from Proposition 7.10 that $\omega(x) \cap L = \{x_0\}$. On the other hand, proceeding as in the proof of Proposition 7.10, we conclude that $\gamma^+(x_k) \subset$

$\omega(x)$ intersects $L$ for any sufficiently large $k$. Since $\omega(x) \cap L = \{x_0\}$, this shows that $x_k \in \gamma(x_0) = \gamma(p)$ for any sufficiently large $k$, which contradicts the choice of the sequence $(x_k)_k$. Therefore, $\omega(x)$ is a periodic orbit.

Finally, we assume that $\omega(x)$ contains noncritical points but no periodic orbits. We show that for any noncritical point $p \in \omega(x)$ the sets $\omega(p)$ and $\alpha(p)$ are critical points. We only consider $\omega(p)$, because the argument for $\alpha(p)$ is analogous. Let $p \in \omega(x)$ be a noncritical point. We note that $\omega(p) \subset \omega(x)$. If $q \in \omega(p)$ is not a critical point and $L$ is transversal to $f$ containing $q$, then, by Proposition 7.10,

$$\omega(x) \cap L = \omega(p) \cap L = \{q\};$$

in particular, the orbit $\gamma^+(p)$ intersects $L$ at a point $x_0$. Since $\gamma^+(p) \subset \omega(x)$, we have $x_0 = q$, and thus $\gamma^+(p)$ and $\omega(p)$ have the point $q$ in common. Proceeding again as in the proof of Proposition 7.10, we conclude that $\omega(p) = \gamma(p)$ is a periodic orbit. This contradiction shows that $\omega(p)$ contains only critical points and since it is connected it contains a single critical point. $\qquad\square$

We recall that by Proposition 7.8 the set $\omega(x)$ is connected. Under the assumptions of Theorem 7.15, this forbids, for example, that $\omega(x)$ is a (finite) union of critical points.

One can also formulate a corresponding result for negative semiorbits.

## 7.3. Exercises

**Exercise 7.1.** Consider the matrices

$$A = \begin{pmatrix} 4 & 1 & 0 & 0 & 0 \\ 0 & 4 & 1 & 0 & 0 \\ 0 & 0 & 4 & 0 & 0 \\ 0 & 0 & 0 & 1 & 1 \\ 0 & 0 & 0 & -1 & 1 \end{pmatrix} \quad \text{and} \quad B = \begin{pmatrix} -1 & 1 & 0 & 0 & 0 \\ 0 & -1 & 0 & 0 & 0 \\ 0 & 0 & 2 & 0 & 0 \\ 0 & 0 & 0 & 0 & -3 \\ 0 & 0 & 0 & 3 & 0 \end{pmatrix}.$$

a) For the equation $x' = Ax$ show that $\alpha(x) = \{0\}$ for every $x \in \mathbb{R}^5$.

b) For the equation $x' = Bx$ show that a solution $x$ is bounded if and only if $x(0) \in \{0\}^3 \times \mathbb{R}^2$.

**Exercise 7.2.** By sketching the phase portrait, verify that there exist equations in the plane with at least one disconnected $\omega$-limit set.

**Exercise 7.3.** Consider the equation

$$\begin{cases} x' = x^2 - xy, \\ y' = y^2 - x^2 - 1. \end{cases}$$

a) Show that the straight line $x = 0$ is a union of orbits.

b) Find whether there exist other straight lines passing through the origin and having the same property.

**Exercise 7.4.** For each $\varepsilon \in \mathbb{R}$, consider the equation in polar coordinates

$$\begin{cases} r' = r(1 - r), \\ \theta' = \sin^2 \theta + \varepsilon. \end{cases}$$

a) Sketch the phase portrait for each $\varepsilon \in \mathbb{R}$.

b) Find all values of $\varepsilon$ for which the equation is conservative.

c) Find the period of each periodic orbit when $\varepsilon = 1$.

d) Find whether the smallest invariant set containing the open ball of radius $1/2$ centered at $(1, 0)$ is an open set when $\varepsilon = 0$.

**Exercise 7.5.** Consider the equation

$$\begin{cases} x' = x^2 - y^2, \\ y' = x^2 + y^2. \end{cases}$$

a) Show that there is an invariant straight line containing $(0, 0)$.

b) Show that there are no periodic orbits.

c) Sketch the phase portrait.

**Exercise 7.6.** For the function $B(x, y) = xy(1 - x - y)$, consider the equation

$$x' = \frac{\partial B(x, y)}{\partial y} \quad \text{and} \quad y' = -\frac{\partial B(x, y)}{\partial x}.$$

a) Find all critical points and verify that the straight lines $x = 0$ and $y = 0$ are invariant.

b) Show that the straight line $x + y = 1$ is invariant.

c) Find an invariant compact set with infinitely many points.

d) Sketch the phase portrait.

**Exercise 7.7.** Given a function $f \colon \mathbb{R}^2 \to \mathbb{R}$ of class $C^2$, consider the equation $x' = \nabla f(x)$. Show that any nonempty $\omega$-limit set is a critical point.

**Exercise 7.8.** Verify that there exists an autonomous equation in $\mathbb{R}^3$ with a periodic orbit but without critical points.

**Solutions.**

**7.3** b) There are none.

**7.4** b) The equation is conservative for no values of $\varepsilon$.

c) The only periodic orbit is the circle of radius 1 centered at the origin and its period is $\int_0^{2\pi} 1/(\sin^2 \theta + 1) \, d\theta = \sqrt{2}\pi$.

d) It is open.

**7.6** a) $(0,0)$, $(0,1)$, $(1,0)$ and $(1/3, 1/3)$.

c) Triangle determined by $(0,0)$, $(0,1)$ and $(1,0)$.

*Part 4*

# Further Topics

# Chapter 8

# Bifurcations and Center Manifolds

This chapter gives a brief introduction to bifurcation theory. We begin with the description of several examples of bifurcations. In particular, among others we consider the Hopf bifurcation, which corresponds to the appearance (or disappearance) of a periodic orbit. We then give an introduction to the theory of center manifolds, which often allows one to reduce the order of an equation in the study of the existence of bifurcations. Center manifolds are also useful in the study of the stability of a critical point, by reducing the problem to the study of the stability on the center manifold. Finally, we give an introduction to the theory of normal forms, which aims to eliminate, through an appropriate change of variables, all possible terms in the original equation. For additional topics we refer the reader to [**8, 12, 14**].

## 8.1. Introduction to bifurcation theory

We start with an example that illustrates the type of problems considered in bifurcation theory.

**Example 8.1.** Consider the equation

$$\begin{cases} x' = x, \\ y' = (1 + \varepsilon)y, \end{cases} \tag{8.1}$$

with $\varepsilon \in \mathbb{R}$. The phase portrait for each value of $\varepsilon$ is shown in Figure 8.1.

We are interested in knowing for which values of $\varepsilon_0 \in \mathbb{R}$ there exists $\varepsilon$ in an arbitrarily small neighborhood of $\varepsilon_0$ such that the solutions of equation (8.1) for $\varepsilon_0$ and $\varepsilon$ are not differentially conjugate (see Definition 2.44).

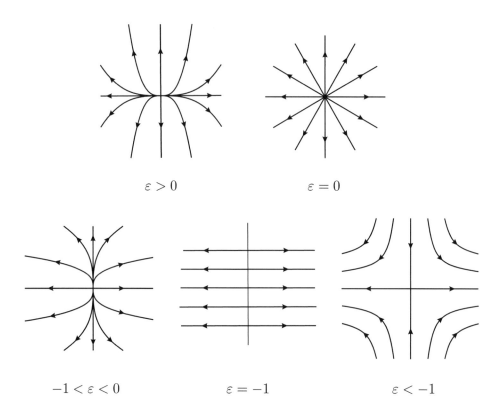

$$\varepsilon > 0 \qquad\qquad\qquad \varepsilon = 0$$

$$-1 < \varepsilon < 0 \qquad\qquad \varepsilon = -1 \qquad\qquad \varepsilon < -1$$

**Figure 8.1.** Phase portrait of equation (8.1) for each $\varepsilon \in \mathbb{R}$.

From the phase portraits in Figure 8.1 it is clear that this happens for $\varepsilon = 0$ and $\varepsilon = -1$. Moreover, it follows from Propositions 2.45 and 2.47 that the same happens for all remaining values of $\varepsilon$, because the matrix

$$\begin{pmatrix} 1 & 0 \\ 0 & 1+\varepsilon \end{pmatrix}$$

of the linear equation (8.1) has different eigenvalues for different values of $\varepsilon$. Hence, the solutions of equation (8.1) for any two different values of $\varepsilon$ are not differentially conjugate. Again, this shows that differentiable conjugacies are somewhat rigid, since they distinguish phase portraits that clearly have the same qualitative behavior (such as the first three phase portraits in Figure 8.1).

Now we consider the analogous problem for the notion of topological conjugacy. In other words, we want to know for which values of $\varepsilon_0 \in \mathbb{R}$ there exists $\varepsilon$ in an arbitrary small neighborhood of $\varepsilon_0$ such that the solutions of equation (8.1) for $\varepsilon_0$ and $\varepsilon$ are not topologically conjugate (see Definition 2.44). In this case, it follows readily from Theorem 2.50 that this only occurs for $\varepsilon = -1$. We then say that a bifurcation occurs in equation (8.1) at $\varepsilon = -1$.

Now we formalize the concept of bifurcation, also allowing topological conjugacies up to a time change along each orbit. Consider a function $f\colon \mathbb{R}^n \times \mathbb{R}^k \to \mathbb{R}^n$ of class $C^1$ and the equation

$$x' = f(x, \varepsilon), \tag{8.2}$$

for each value of the parameter $\varepsilon \in \mathbb{R}^k$.

**Definition 8.2.** We say that a *bifurcation* does not occur in equation (8.2) at $\varepsilon = \varepsilon_0$ if for each arbitrarily close $\varepsilon_1 \in \mathbb{R}^k$ there exist a homeomorphism $h\colon \mathbb{R}^n \to \mathbb{R}^n$ and a continuous function $\tau\colon \mathbb{R} \times \mathbb{R}^n \to \mathbb{R}$ with $t \mapsto \tau(t, x)$ increasing for each $x \in \mathbb{R}^n$ such that

$$h(\varphi_t(x)) = \psi_{\tau(t,x)}(h(x)) \tag{8.3}$$

for every $t \in \mathbb{R}$ and $x \in \mathbb{R}^n$, where $\varphi_t(z)$ and $\psi_t(z)$ are, respectively, the solutions of the initial value problems

$$\begin{cases} x' = f(x, \varepsilon_0), \\ x(0) = z \end{cases} \quad \text{and} \quad \begin{cases} x' = f(x, \varepsilon_1), \\ x(0) = z. \end{cases} \tag{8.4}$$

In other words, a bifurcation occurs at $\varepsilon = \varepsilon_0$ if in any arbitrarily small neighborhood of $\varepsilon_0$ there exists $\varepsilon_1$ such that the solutions of the equations

$$x' = f(x, \varepsilon_0) \quad \text{and} \quad x' = f(x, \varepsilon_1)$$

are not transformed into each other by a homeomorphism preserving orientation.

**Example 8.3.** Consider the equation

$$\begin{cases} x' = (1 + \varepsilon^2)y, \\ y' = -(1 + \varepsilon^2)x. \end{cases} \tag{8.5}$$

The origin is a critical point and the remaining orbits are circular periodic orbits centered at the origin and of period $2\pi/(1+\varepsilon^2)$. Indeed, equation (8.3) can be written in polar coordinates in the form

$$\begin{cases} r' = 0, \\ \theta' = -(1 + \varepsilon^2). \end{cases}$$

Since

$$\varphi_t = \psi_{t(1+\varepsilon_0^2)/(1+\varepsilon_1^2)},$$

with $\varphi_t$ and $\psi_t$ defined by (8.4), taking

$$h(x) = x \quad \text{and} \quad \tau(t) = t(1 + \varepsilon_0^2)/(1 + \varepsilon_1^2)$$

yields identity (8.3). This shows that no bifurcations occur in equation (8.5).

**Example 8.4.** Consider the equation

$$x' = \varepsilon x - x^2. \tag{8.6}$$

Clearly, $x = 0$ and $x = \varepsilon$ are critical points. The phase portrait is the one shown in Figure 8.2. One can easily verify that the only bifurcation occurs at $\varepsilon = 0$. It is called a *transcritical bifurcation* and corresponds to the collision of two critical points, one stable and one unstable, that exchange their stability after the collision.

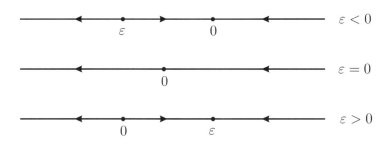

**Figure 8.2.** Phase portrait of equation (8.6).

**Example 8.5.** Consider the equation

$$x' = \varepsilon - x^2. \tag{8.7}$$

The number of critical points depends on the sign of $\varepsilon$. For $\varepsilon < 0$ there are no critical points, for $\varepsilon = 0$ the origin is the only critical point, and finally, for $\varepsilon > 0$ there are two critical points, namely $-\sqrt{\varepsilon}$ and $\sqrt{\varepsilon}$. The phase portrait is the one shown in Figure 8.3. Clearly, the only bifurcation in equation (8.7) occurs at $\varepsilon = 0$. It is called a *saddle-node bifurcation* (see also Example 8.6) and corresponds to the collision of two critical points, one stable and one unstable, that are annihilated after the collision.

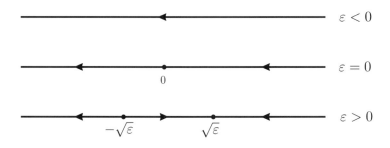

**Figure 8.3.** Phase portrait of equation (8.7).

**Example 8.6.** Consider the equation

$$\begin{cases} x' = \varepsilon - x^2, \\ y' = y. \end{cases} \tag{8.8}$$

We note that the first component coincides with equation (8.7), and thus, the phase portrait of equation (8.8) is the one shown in Figure 8.4. Again, the only bifurcation occurs at $\varepsilon = 0$. It can be described as the collision of a saddle point and a node for $\varepsilon > 0$, which disappear for $\varepsilon < 0$. This justifies the name of the bifurcation in Example 8.5.

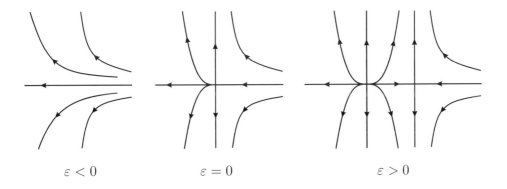

$\varepsilon < 0$        $\varepsilon = 0$        $\varepsilon > 0$

**Figure 8.4.** Phase portrait of equation (8.8).

**Example 8.7.** Consider the equation

$$x' = \varepsilon x - x^3. \tag{8.9}$$

For $\varepsilon \leq 0$ the origin is the only critical point, while for $\varepsilon > 0$ there are two critical points, namely $-\sqrt{\varepsilon}$ and $\sqrt{\varepsilon}$. The phase portrait is the one shown in Figure 8.5. One can easily verify that the only bifurcation occurs at $\varepsilon = 0$. It is called a *pitchfork bifurcation* and corresponds to the creation (or annihilation) of two critical points, one stable and one unstable.

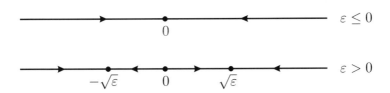

**Figure 8.5.** Phase portrait of equation (8.9).

**Example 8.8.** Consider the equation

$$\begin{cases} x' = \varepsilon x - y - x(x^2 + y^2), \\ y' = x + \varepsilon y - y(x^2 + y^2), \end{cases} \tag{8.10}$$

which in polar coordinates takes the form

$$\begin{cases} r' = \varepsilon r - r^3, \\ \theta' = 1. \end{cases} \tag{8.11}$$

We note that the first component in (8.11) was already considered in Example 8.7 (although now we are only interested in the nonnegative values of the variable, because $r \geq 0$). The phase portrait of equation (8.10), or of equation (8.11), is the one shown in Figure 8.6. One can easily verify that the only bifurcation in equation (8.10) occurs at $\varepsilon = 0$. It is called a *Hopf bifurcation* and corresponds to the creation (or annihilation) of a periodic orbit.

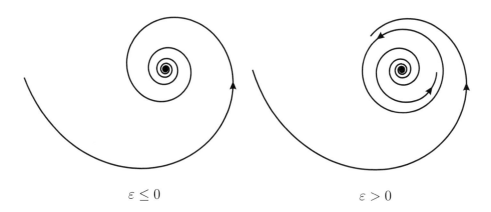

$$\varepsilon \leq 0 \qquad\qquad\qquad\qquad\qquad \varepsilon > 0$$

**Figure 8.6.** Phase portrait of equation (8.10).

There are many other bifurcations, but a systematic study of bifurcation theory clearly falls outside the scope of the book (for detailed treatments we refer the reader to [**8, 12**]).

## 8.2. Center manifolds and applications

In this section we give a brief introduction to the theory of center manifolds, and we illustrate how it can be of help in bifurcation theory.

**8.2.1. Basic notions.** Let $f \colon \mathbb{R}^n \to \mathbb{R}^n$ be a function of class $C^1$ and let $x_0$ be a critical point of the equation $x' = f(x)$. Unlike what happens in Section 4.1, here we do not assume that $x_0$ is a hyperbolic critical point. We continue to write $A = d_{x_0} f$.

**Definition 8.9.** We define the *stable*, *unstable*, and *center spaces* of $x_0$, respectively, by

$$E^s = \left\{ x \in \mathbb{R}^n \setminus \{0\} : \limsup_{t \to +\infty} \frac{1}{t} \log \|e^{At} x\| < 0 \right\} \cup \{0\},$$

$$E^u = \left\{ x \in \mathbb{R}^n \setminus \{0\} : \limsup_{t \to -\infty} \frac{1}{|t|} \log \|e^{At} x\| < 0 \right\} \cup \{0\},$$

$$E^c = \left\{ x \in \mathbb{R}^n \setminus \{0\} : \limsup_{t \to \pm\infty} \frac{1}{|t|} \log \|e^{At} x\| = 0 \right\} \cup \{0\}.$$

In other words, $E^s$ and $E^u$ contain the initial conditions (other than the origin) whose solutions have some exponential behavior, while $E^c$ contains the initial conditions whose solutions have no exponential behavior. One can easily verify that in the case of a hyperbolic critical point the sets $E^s$ and $E^u$ in Definition 8.9 coincide with the stable and unstable spaces introduced in Definition 4.2.

**Proposition 8.10.** *If $x_0$ is a critical point of the equation $x' = f(x)$, then:*

    a) *$E^s$, $E^u$ and $E^c$ are subspaces of $\mathbb{R}^n$ and $E^s \oplus E^u \oplus E^c = \mathbb{R}^n$;*

    b) *for every $t \in \mathbb{R}$ we have*

$$e^{At}(E^s) \subset E^s, \quad e^{At}(E^u) \subset E^u \quad \text{and} \quad e^{At}(E^c) \subset E^c.$$

**Proof.** One can proceed in a similar manner to that in the proof of Proposition 4.3. Namely, the Jordan canonical form of the matrix $A = d_{x_0} f$ can be written in the block form

$$\begin{pmatrix} A_s & 0 & 0 \\ 0 & A_u & 0 \\ 0 & 0 & A_c \end{pmatrix}$$

with respect to the decomposition $\mathbb{R}^n = E^s \oplus E^u \oplus E^c$. The matrices $A_s$, $A_u$ and $A_c$ correspond, respectively, to the Jordan blocks of eigenvalues with negative, positive and zero real part. $\qquad\square$

When $E^c = \{0\}$, that is, when the critical point $x_0$ is hyperbolic, the Grobman–Hartman theorem and the Hadamard–Perron theorem (Theorems 4.7 and 5.2) describe with sufficient detail the phase portrait of the equation $x' = f(x)$ in a neighborhood of $x_0$. In particular, the solutions of the equations $x' = f(x)$ and $y' = Ay$ are topologically conjugate, respectively, in neighborhoods of $x_0$ and $0$. Moreover, there exist invariant manifolds (in the sense that we have the inclusions in (5.6)) that contain $x_0$ and are tangent, respectively, to the stable and unstable spaces $E^s$ and $E^u$.

In addition, Theorem 4.12 shows that any sufficiently small $C^1$ perturbation of a $C^1$ vector field with a hyperbolic critical point has a homeomorphic phase portrait. Therefore, there are no bifurcations in the neighborhood of a hyperbolic critical point under sufficiently small $C^1$ perturbations. Hence, bifurcations may only occur when $E^c \neq \{0\}$.

**8.2.2. Center manifolds.** We start the discussion of the nonhyperbolic case (when $E^c \neq \{0\}$) with a result that is analogous to the Hadamard–Perron theorem (Theorem 5.2).

**Theorem 8.11** (Center manifold theorem). *If $x_0 \in \mathbb{R}^n$ is a critical point of the equation $x' = f(x)$ for a function $f \colon \mathbb{R}^n \to \mathbb{R}^n$ of class $C^k$, with $k \in \mathbb{N}$, then there exist manifolds $W^s$, $W^u$ and $W^c$ of class $C^k$ containing $x_0$ such that:*

  a) *$T_{x_0} W^s = E^s$, $T_{x_0} W^u = E^u$ and $T_{x_0} W^c = E^c$;*

  b) *the solutions of the equation $x' = f(x)$ with initial condition in $W^s$, $W^u$ and $W^c$ remain in these manifolds for any sufficiently small time.*

*Moreover, the manifolds $W^s$ and $W^u$ are uniquely determined by these properties in any sufficiently small neighborhood of $x_0$.*

The proof of Theorem 8.11 falls outside the scope of the book (see [**6**] for details). However, it corresponds to an elaboration of the proof of the Hadamard–Perron theorem (Theorem 5.2). We note that Theorem 8.11 contains Theorem 5.2 as a particular case (when $x_0$ is a hyperbolic critical point).

**Definition 8.12.** The manifolds $W^s$, $W^u$ and $W^c$ are called, respectively, *stable*, *unstable*, and *center manifolds*.

Unlike what happens with the stable and unstable manifolds, the center manifold $W^c$ in Theorem 8.11 may not be unique.

**Example 8.13.** Consider the equation

$$\begin{cases} x' = x^2, \\ y' = -y. \end{cases} \tag{8.12}$$

The origin is a critical point and for the function $f(x, y) = (x^2, -y)$ we have

$$d_{(0,0)} f = \begin{pmatrix} 0 & 0 \\ 0 & -1 \end{pmatrix}.$$

Thus,

$$E^s = \{0\} \times \mathbb{R}, \quad E^u = \{0\} \quad \text{and} \quad E^c = \mathbb{R} \times \{0\}.$$

One can easily verify that for the initial condition $(x(0), y(0)) = (x_0, y_0)$ equation (8.12) has the solution

$$(x(t), y(t)) = \left( \frac{x_0}{1 - tx_0}, e^{-t} y_0 \right).$$

Eliminating $t$ we obtain $y(x) = (y_0 e^{-1/x_0}) e^{1/x}$, which yields the phase portrait in Figure 8.7.

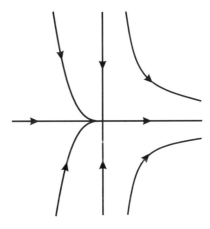

**Figure 8.7.** Phase portrait of equation (8.12).

Since each orbit in the half-plane $\mathbb{R}^- \times \mathbb{R}$ is tangent to the horizontal axis, there are infinitely many center manifolds, given by

$$W^c = \{(x, \eta_c(x)) : x \in \mathbb{R}\}$$

where

$$\eta_c(x) = \begin{cases} ce^{1/x} & \text{if } x < 0, \\ 0 & \text{if } x \geq 0 \end{cases}$$

for an arbitrary constant $c \in \mathbb{R}$. One can easily verify that each manifold $W^c$ is of class $C^\infty$ (but is not analytic).

**8.2.3. Applications of center manifolds.** We describe briefly in this section how the center manifolds given by Theorem 8.11 can be used in the study of the stability of a (nonhyperbolic) critical point.

Consider variables $\bar{x}, \bar{y}, \bar{z}$ parameterizing $W^s \times W^u \times W^c$ in a sufficiently small neighborhood of $x_0$. One can show, with an appropriate generalization of the Grobman–Hartman theorem together with Theorem 2.50, that the solutions of the equation $x' = f(x)$ are topologically conjugate to those of

$$\begin{cases} \bar{x}' = -\bar{x}, \\ \bar{y}' = \bar{y}, \\ \bar{z}' = F(\bar{z}) \end{cases} \tag{8.13}$$

for some function $F$ (see [6] for details). When $E^u \neq \{0\}$ it follows from (8.13) that the critical point $x_0$ is unstable. Now we assume that $E^u = \{0\}$. In this case, if $E^c = \{0\}$, then $x_0$ is asymptotically stable. On the other hand, if $E^u = \{0\}$ but $E^c \neq \{0\}$, then the stability of $x_0$ coincides with the stability of the origin in the equation

$$\bar{z}' = F(\bar{z}) \tag{8.14}$$

(assuming that $x_0$ is represented by $\bar{z} = 0$). In summary, we have three cases:

   a) if $E^u \neq \{0\}$, then $x_0$ is unstable;
   b) if $E^u = \{0\}$ and $E^c = \{0\}$, then $x_0$ is asymptotically stable;
   c) if $E^u = \{0\}$ and $E^c \neq \{0\}$, then the stability of $x_0$ coincides with the stability of the origin in equation (8.14).

In the third case, it is sufficient to study the behavior on the center manifold.

**Example 8.14.** Consider the equation

$$\begin{cases} x' = -x + y^2, \\ y' = y^2 - x^2. \end{cases} \tag{8.15}$$

The origin is a critical point and for the function

$$f(x, y) = (-x + y^2, y^2 - x^2)$$

we have

$$d_{(0,0)}f = \begin{pmatrix} -1 & 0 \\ 0 & 0 \end{pmatrix}.$$

Thus

$$E^s = \mathbb{R} \times \{0\}, \quad E^u = \{0\} \quad \text{and} \quad E^c = \{0\} \times \mathbb{R},$$

and we are in the third case. Since $f$ is of class $C^\infty$, it follows from the Center manifold theorem (Theorem 8.11) that there exist manifolds $W^s$ and $W^c$ of class $C^k$ for any $k \in \mathbb{N}$. Moreover,

$$W^c = \big\{ (\varphi(y), y) : y \in (-\delta, \delta) \big\}$$

for some function $\varphi \colon (-\delta, \delta) \to \mathbb{R}$ of class $C^k$ with $\varphi(0) = \varphi'(0) = 0$, because $0 \in W^c$ and $T_0 W^c = E^c$. Thus, one can write the Taylor series

$$\varphi(y) = ay^2 + by^3 + \cdots \tag{8.16}$$

(up to order $k$). Substituting $x = \varphi(y)$ in (8.15), we obtain

$$x' = \varphi'(y)y' = \varphi'(y)(y^2 - \varphi(y)^2),$$

or equivalently,

$$-\varphi(y) + y^2 = \varphi'(y)(y^2 - \varphi(y)^2).$$

Using (8.16) yields the identity

$$-ay^2 - by^3 - \cdots + y^2 = (2ay + 3by^2 + \cdots)(y^2 - a^2y^4 - \cdots).$$

Equating terms of the same order in this identity, we obtain

$$-a + 1 = 0, \quad -b = 2a, \ldots,$$

that is,

$$a = 1, \quad b = -2, \ldots.$$

Hence,

$$\varphi(y) = y^2 - 2y^3 + \cdots, \tag{8.17}$$

and it follows again from (8.15) that the equation $\bar{z}' = F(\bar{z})$ takes the form

$$y' = y^2 - \varphi(y)^2 = y^2 - y^4 - \cdots. \tag{8.18}$$

In a sufficiently small neighborhood of $y = 0$ the sign of $y'$ coincides with the sign of $y^2$, and thus the origin is unstable in (8.18). According to the previous discussion, this shows that the origin is unstable in equation (8.15). Moreover, it follows from (8.17) and (8.18) that in a sufficiently small neighborhood of the origin the center manifold $W^c$ and the motion along $W^c$ are those shown in Figure 8.8.

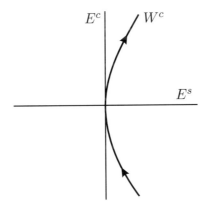

**Figure 8.8.** Center manifold of equation (8.15).

Although it is not necessary for the study of the stability of the origin, one can also use the former procedure to approximate the stable manifold. Namely, we have

$$W^s = \{(x, \psi(x)) : x \in (-\delta, \delta)\}$$

for some function $\psi \colon (-\delta, \delta) \to \mathbb{R}$ of class $C^k$ with $\psi(0) = \psi'(0) = 0$, because $0 \in W^s$ and $T_0 W^s = E^s$. Then one can write

$$\psi(x) = \alpha x^2 + \beta x^3 + \cdots. \tag{8.19}$$

Substituting $y = \psi(x)$ in (8.15), we obtain

$$y' = \psi'(x)x' = \psi'(x)(-x + \psi(x)^2),$$

and thus,

$$\psi(x)^2 - x^2 = \psi'(x)(-x + \psi(x)^2).$$

Using (8.19) yields the identity

$$\alpha^2 x^4 + \cdots - x^2 = (2\alpha x + 3\beta x^2 + \cdots)(-x + \alpha^2 x^4 + \cdots).$$

Equating terms in this identity, we obtain

$$-1 = -2\alpha, \quad 0 = -3\beta, \ldots,$$

and thus,

$$\psi(x) = \frac{1}{2}x^2 + 0x^3 + \cdots.$$

This shows that in a neighborhood of the origin the phase portrait is the one shown in Figure 8.9.

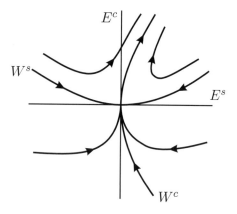

**Figure 8.9.** Phase portrait of equation (8.15) in a neighborhood of the origin.

**Example 8.15.** Consider the equation

$$\begin{cases} x' = x^2 - xy, \\ y' = -y + x^2. \end{cases} \tag{8.20}$$

The origin is a critical point and for the function

$$f(x, y) = (x^2 - xy, -y + x^2)$$

we have

$$d_{(0,0)}f = \begin{pmatrix} 0 & 0 \\ 0 & -1 \end{pmatrix}.$$

Hence,

$$E^s = \{0\} \times \mathbb{R}, \quad E^u = \{0\} \quad \text{and} \quad E^c = \mathbb{R} \times \{0\}.$$

Since $f(0, y) = (0, -y)$, the function $f$ is vertical on the vertical axis. This implies that the vertical axis is invariant. By the uniqueness of the stable manifold in Theorem 8.11, we conclude that $W^s = E^s$.

Now we consider the center manifold

$$W^c = \{(x, \psi(x)) : x \in (-\delta, \delta)\},$$

where

$$\psi(x) = ax^2 + bx^3 + \cdots . \tag{8.21}$$

Substituting $y = \psi(x)$ in equation (8.20), we obtain

$$y' = \psi'(x)x' = \psi'(x)(x^2 - x\psi(x)),$$

and thus,

$$-\psi(x) + x^2 = \psi'(x)(x^2 - x\psi(x)).$$

Using (8.21) yields the identity

$$-ax^2 - bx^3 - \cdots + x^2 = (2ax + 3bx^2 + \cdots)(x^2 - ax^3 - bx^4 - \cdots),$$

and we conclude that

$$a = 1, \quad 2a = -b, \ldots.$$

Thus,

$$\psi(x) = x^2 - 2x^3 + \cdots,$$

and

$$x' = x^2 - x\psi(x) = x^2 - x^3 + 2x^4 + \cdots.$$

Therefore, the origin is unstable in equation (8.20). In a neighborhood of the origin the phase portrait is the one shown in Figure 8.10.

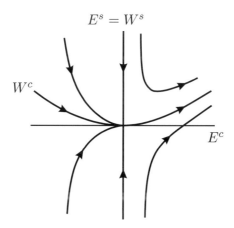

**Figure 8.10.** Phase portrait of equation (8.20) in a neighborhood of the origin.

Now we illustrate with an example how center manifolds can be used to simplify the study of bifurcation theory.

**Example 8.16.** Consider the equation

$$\begin{cases} x' = \varepsilon x - x^3 + y^2, \\ y' = -y + x^2. \end{cases} \tag{8.22}$$

Adding a third component $\varepsilon' = 0$, we obtain

$$\begin{pmatrix} x \\ y \\ \varepsilon \end{pmatrix}' = \begin{pmatrix} 0 & 0 & 0 \\ 0 & -1 & 0 \\ 0 & 0 & 0 \end{pmatrix} \begin{pmatrix} x \\ y \\ \varepsilon \end{pmatrix} + \begin{pmatrix} \varepsilon x - x^3 + y^2 \\ x^2 \\ 0 \end{pmatrix}. \tag{8.23}$$

Since the $3 \times 3$ matrix in (8.23) has eigenvalues 0, 0 and $-1$, it follows from the Center manifold theorem (Theorem 8.11) that there exist manifolds $W^s$ and $W^c$ that are tangent respectively to the stable and center spaces

$$E^s = \{0\} \times \mathbb{R} \times \{0\} \quad \text{and} \quad E^c = \mathbb{R} \times \{0\} \times \mathbb{R}.$$

It follows from the discussion in the beginning of this section that in order to determine the stability of the origin in (8.23) (and thus in (8.22)) it suffices to study the behavior on the center manifold (or more precisely, on any center manifold). It can be written in the form

$$W^c = \{(x, \varphi(x, \varepsilon), \varepsilon) : x, \varepsilon \in (-\delta, \delta)\}$$

for some function $\varphi \colon (-\delta, \delta) \times (-\delta, \delta) \to \mathbb{R}$ with $\varphi(0,0) = 0$ and $d_{(0,0)}\varphi = 0$. For $y = \varphi(x, \varepsilon)$ we have

$$y' = \frac{\partial \varphi}{\partial x} x' + \frac{\partial \varphi}{\partial \varepsilon} \varepsilon' = \frac{\partial \varphi}{\partial x} (\varepsilon x - x^3 + y^2),$$

and by (8.22), we obtain

$$\frac{\partial \varphi}{\partial x} (\varepsilon x - x^3 + y^2) = -\varphi(x, \varepsilon) + x^2. \tag{8.24}$$

Writing

$$\varphi(x, \varepsilon) = ax^2 + bx\varepsilon + c\varepsilon^2 + \cdots,$$

it follows from (8.24) that

$$(2ax + b\varepsilon + \cdots)(\varepsilon x - x^3 + \varphi(x, \varepsilon)^2) = -ax^2 - bx\varepsilon - c\varepsilon^2 - \cdots + x^2,$$

which yields $a = b = c = 0$. Hence, by the first equation in (8.22) we obtain

$$x' = \varepsilon x - x^3 + \varphi(x, \varepsilon)^2 = \varepsilon x - x^3 + \cdots.$$

The behavior of the solutions of equation (8.22) in a neighborhood of the origin can now be deduced from the equation $x' = \varepsilon x - x^3$, which was already studied in Example 8.7.

## 8.3. Theory of normal forms

As noted earlier, there are many types of bifurcations and it is desirable to have some kind of classification. While a corresponding systematic study falls outside the scope of the book, it is at least convenient to have a procedure, as automatic as possible, that can simplify a large class of equations before we study the existence of bifurcations. The theory of normal forms, which we introduce in this section, provides such a procedure.

We begin with an example.

**Example 8.17.** For each $\varepsilon \in \mathbb{R}$, consider the scalar equation

$$x' = \varepsilon x + f(x),$$

where $f\colon \mathbb{R} \to \mathbb{R}$ is a function of class $C^\infty$ with $f(0) = f'(0) = 0$. Writing the Taylor series of $f$, we obtain

$$x' = \varepsilon x + ax^2 + bx^3 + \cdots. \tag{8.25}$$

Now we consider a change of variables of the form

$$x = y + \alpha y^2 + \beta y^3 + \cdots. \tag{8.26}$$

If the series in (8.26) has a positive radius of convergence (in particular, if it is a polynomial), then indeed it defines a change of variables in a sufficiently small neighborhood of the origin. Substituting (8.26) in (8.25), we obtain

$$y'(1 + 2\alpha y + 3\beta y^2 + \cdots) = \varepsilon(y + \alpha y^2 + \beta y^3) + a(y^2 + 2\alpha y^3) + by^3 + \cdots,$$

and thus,

$$y' = \frac{\varepsilon y + (\varepsilon\alpha + a)y^2 + \varepsilon\beta + (2a\alpha + b)y^3 + \cdots}{1 + 2\alpha y + 3\beta y^2 + \cdots}.$$

Using the identity

$$\sum_{k=0}^{\infty} (-r)^k = \frac{1}{1+r},$$

which holds for $|r| < 1$, after some computations we obtain

$$
\begin{aligned}
y' &= \big(\varepsilon y + (\varepsilon\alpha + a)y^2 + \cdots\big) \\
&\quad \times \big[1 - 2\alpha y - 3\beta y^2 - \cdots + (2\alpha y + 3\beta y^2 + \cdots)^2 - \cdots\big] \\
&= \varepsilon y + (a - \varepsilon\alpha)y^2 + (b + 2\varepsilon\alpha^2 - 2\varepsilon\beta)y^3 + \cdots.
\end{aligned} \tag{8.27}
$$

In particular, choosing the coefficients $\alpha$ and $\beta$ so that the coefficients of $y^2$ and $y^3$ in (8.27) vanish, that is, taking

$$\alpha = \frac{a}{\varepsilon} \quad \text{and} \quad \beta = \frac{a^2}{\varepsilon^2} + \frac{b}{2\varepsilon}$$

for $\varepsilon \neq 0$, we obtain the equation

$$y' = \varepsilon y + o(y^3).$$

LIVERPOOL JOHN MOORES UNIVERSITY
LEARNING SERVICES

Now we consider the general case. Given a matrix $A \in M_n$ and a function $f \colon \mathbb{R}^n \to \mathbb{R}^n$ of class $C^\infty$ with $f(0) = 0$ and $d_0 f = 0$, consider the equation

$$x' = Ax + f(x).$$

In a similar manner to that in Example 8.17, we consider a change of variables $x = y + h(y)$, where $h \colon \mathbb{R}^n \to \mathbb{R}^n$ is a differentiable function with $h(0) = 0$ and $d_0 h = 0$. Then

$$x' - y' + d_y h y' = (\mathrm{Id} + d_y h) y'.$$

Now we recall that

$$\sum_{k=0}^{\infty} (-B)^k = (\mathrm{Id} + B)^{-1} \tag{8.28}$$

for any square matrix $B$ with $\|B\| < 1$. Since $d_0 h = 0$, it follows from (8.28) that for any sufficiently small $y$ the matrix $\mathrm{Id} + d_y h$ is invertible, and one can write

$$y' = (\mathrm{Id} + d_y h)^{-1} x'$$
$$= (\mathrm{Id} + d_y h)^{-1} \big[ A(y + h(y)) + f(y + h(y)) \big].$$

It also follows from (8.28) that

$$y' = (\mathrm{Id} - d_y h + \cdots) \big[ Ay + Ah(y) + f(y + h(y)) \big]. \tag{8.29}$$

Now let $H_{m,n}$ be the set of all functions $p \colon \mathbb{R}^n \to \mathbb{R}^n$ such that each component is a homogeneous polynomial of degree $m$, that is, a linear combination of polynomials of the form $x_{i_1}^{m_1} \cdots x_{i_k}^{m_k}$ with

$$k \in \mathbb{N}, \quad i_1, \ldots, i_k \in \{1, \ldots, n\} \quad \text{and} \quad m_1 + \cdots + m_k = m.$$

We note that $H_{m,n}$ is a linear space. Write

$$f = f_2 + f_3 + \cdots \tag{8.30}$$

and

$$h = \mathrm{Id} + h_2 + h_3 + \cdots,$$

where $f_m, g_m \in H_{m,n}$ for each $m \geq 2$. It follows from (8.29) that

$$y' = Ay + [Ah_2(y) + f_2(y) - D_y h_2 Ay]$$
$$+ \sum_{k=3}^{+\infty} [Ah_k(y) + g_k(y) - D_y h_k Ay], \tag{8.31}$$

where each function $g_k \colon \mathbb{R}^n \to \mathbb{R}^n$ depends only on $h_2, \ldots, h_{k-1}$. In order to eliminate all possible terms in the right-hand side of (8.31), we try to solve, recursively, the equations

$$D_y h_2 Ay - Ah_2(y) = f_2(y) \tag{8.32}$$

and

$$D_y h_k Ay - Ah_k(y) = g_k(y) \tag{8.33}$$

for $k \geq 3$, finding successively functions $h_2, h_3, \ldots$. Equations (8.32) and (8.33) are called *homological equations*.

For each $m \geq 2$ and $n \in \mathbb{N}$, we define a map

$$L_A^{m,n} \colon H_{m,n} \to H_{m,n}$$

by

$$(L_A^{m,n} h)(y) = D_y h A y - A h(y).$$

One can easily verify that $L_A^{m,n}$ is a linear transformation. The homological equations can be written in the form

$$L_A^{2,n} h_2 = f_2 \quad \text{and} \quad L_A^{k,n} h_k = g_k, \ k \geq 3. \qquad (8.34)$$

**Example 8.18.** Let $m = n = 2$ and $A = \left( \begin{smallmatrix} 0 & 1 \\ -1 & 0 \end{smallmatrix} \right)$. Each function $h \in H_{2,2}$ can be written in the form

$$h(x,y) = \left( ax^2 + bxy + cy^2, dx^2 + exy + fy^2 \right),$$

for some constants $a, b, c, d, e$ and $f$. In particular, $\dim H_{2,2} = 6$. We obtain

$$
\begin{aligned}
(L_A^{2,2} h)(x,y) &= \begin{pmatrix} 2ax + by & bx + 2cy \\ 2dx + ey & ex + 2fy \end{pmatrix} \begin{pmatrix} y \\ -x \end{pmatrix} - \begin{pmatrix} dx^2 + exy + fy^2 \\ -ax^2 - bxy - cy^2 \end{pmatrix} \\
&= a \begin{pmatrix} 2xy \\ x^2 \end{pmatrix} + b \begin{pmatrix} y^2 - x^2 \\ xy \end{pmatrix} + c \begin{pmatrix} -2xy \\ y^2 \end{pmatrix} \\
&\quad + d \begin{pmatrix} -x^2 \\ 2xy \end{pmatrix} + e \begin{pmatrix} -xy \\ y^2 - x^2 \end{pmatrix} + f \begin{pmatrix} -y^2 \\ -2xy \end{pmatrix},
\end{aligned}
$$

and thus, the linear transformation $L_A^{2,2}$ is represented by the matrix

$$
\begin{pmatrix}
0 & -1 & 0 & -1 & 0 & 0 \\
2 & 0 & -2 & 0 & -1 & 0 \\
0 & -1 & 0 & 0 & 0 & -1 \\
1 & 0 & 0 & 0 & -1 & 0 \\
0 & 1 & 0 & 2 & 0 & -2 \\
0 & 0 & 1 & 0 & 1 & 0
\end{pmatrix}
$$

with respect to the basis of $H_{2,2}$ given by

$$\left\{ \begin{pmatrix} x^2 \\ 0 \end{pmatrix}, \begin{pmatrix} xy \\ 0 \end{pmatrix}, \begin{pmatrix} y^2 \\ 0 \end{pmatrix}, \begin{pmatrix} 0 \\ x^2 \end{pmatrix}, \begin{pmatrix} 0 \\ xy \end{pmatrix}, \begin{pmatrix} 0 \\ y^2 \end{pmatrix} \right\}.$$

We note that in order to solve the homological equations in (8.34) it is sufficient to find the inverse of each transformation $L_A^{m,n}$, when it exists. But since $L_A^{m,n}$ is linear, in order to decide whether the transformation is invertible it is sufficient to compute its eigenvalues.

**Proposition 8.19.** *If $\lambda_1, \ldots, \lambda_n$ are the eigenvalues of $A$ (counted with their multiplicities), then the eigenvalues of $L_A^{m,n}$ are the numbers*

$$-\lambda_k + \sum_{i=1}^{n} m_i \lambda_i, \tag{8.35}$$

*with*

$$k \in \{1, \ldots, n\}, \quad m_1, \ldots, m_n \in \mathbb{N} \cup \{0\} \quad and \quad m_1 + \cdots + m_n = m.$$

*Moreover, the eigenvectors of $L_A^{m,n}$ associated to the eigenvalue in (8.35) are the polynomials*

$$x_1^{m_1} \cdots x_n^{m_n} v_k, \tag{8.36}$$

*where $v_k$ is an eigenvector of $A$ associated to the eigenvalue $\lambda_k$.*

The following is a particular case.

**Example 8.20.** For the diagonal matrix

$$A = \begin{pmatrix} \lambda_1 & & 0 \\ & \ddots & \\ 0 & & \lambda_n \end{pmatrix},$$

if $e_1, \ldots, e_n$ is the standard basis of $\mathbb{R}^n$, then

$$L_A^{m,n}(x_1^{m_1} \cdots x_n^{m_n} e_k)(x_1, \ldots, x_n)$$

$$= \begin{pmatrix} 0 \\ \vdots \\ 0 \\ (m_1 x_1^{m_1-1} x_2 \cdots x_n) \cdots (m_n x_1^{m_1} \cdots x_n^{m_n-1}) \\ 0 \\ \vdots \\ 0 \end{pmatrix} \begin{pmatrix} \lambda_1 x_1 \\ \vdots \\ \lambda_n x_n \end{pmatrix}$$

$$\quad - x_1^{m_1} \cdots x_n^{m_n} \lambda_k e_k$$

$$= (m_1 \lambda_1 + \cdots + m_n \lambda_n) x_1^{m_1} \cdots x_n^{m_n} e_k - x_1^{m_1} \cdots x_n^{m_n} \lambda_k e_k$$

$$= \left( \sum_{i=1}^{n} m_i \lambda_i - \lambda_k \right) x_1^{m_1} \cdots x_n^{m_n} e_k.$$

This shows that $x_1^{m_1} \cdots x_n^{m_n} e_k \in H_{m,n}$ is an eigenvector of $L_A^{m,n}$ associated to the eigenvalue in (8.35).

By Proposition 8.19, one may not be able to solve a homological equation in (8.34) if some of the numbers in (8.35) are zero. This motivates the following notion.

**Definition 8.21.** The vector $(\lambda_1, \ldots, \lambda_n)$ is said to be *resonant* if there exist integers $k \in \{1, \ldots, n\}$ and $m_1, \ldots, m_n \in \mathbb{N} \cup \{0\}$, with $m_1 + \cdots + m_n \geq 2$, such that

$$\lambda_k = \sum_{i=1}^{n} m_i \lambda_i.$$

When there are no resonant vectors one can solve all homological equations and thus eliminate all terms $f_m$ in (8.30). We recall that for a function $g \colon \mathbb{R}^n \to \mathbb{R}^n$ we write $g(x) = o(\|x\|^k)$ if $g(x)/\|x\|^k \to 0$ when $x \to 0$.

**Theorem 8.22** (Poincaré). *Let $A$ be an $n \times n$ matrix and let $f \colon \mathbb{R}^n \to \mathbb{R}^n$ be a function of class $C^\infty$ with $f(0) = 0$ and $d_0 f = 0$. If the vector formed by the eigenvalues of $A$ is not resonant, then for each $k \in \mathbb{N}$ the equation $x' = Ax + f(x)$ can be transformed into*

$$y' = Ay + o(\|y\|^k)$$

*by a change of variables $x = y + h(y)$.*

**Proof.** Since the vector $(\lambda_1, \ldots, \lambda_n)$ formed by the eigenvalues of $A$ is not resonant, all the eigenvalues of $L_A^{m,n}$ in (8.35) are different from zero. Thus, each linear transformation $L_A^{m,n}$ is invertible and one can obtain successively solutions of all homological equations in (8.34) up to order $k$. $\qquad \square$

Now we describe briefly what happens when there are resonant vectors. In this case one can only invert the restriction of $L_A^{m,n}$ to the subspace $E$ of $H_{m,n}$ generated by the root spaces corresponding to the nonzero eigenvalues in (8.35). More precisely, consider the decomposition $H_{m,n} = E \oplus F$, where $F$ is the subspace of $H_{m,n}$ generated by the root spaces corresponding to the zero eigenvalues in (8.35). Since

$$L_A^{m,n} E = E \quad \text{and} \quad L_A^{m,n} F = F,$$

one can write the linear transformation $L_A^{m,n}$ in the form

$$\begin{pmatrix} L_E & 0 \\ 0 & L_F \end{pmatrix}$$

with respect to the decomposition $H_{m,n} = E \oplus F$. We also write

$$f_2 = \begin{pmatrix} f_2^E \\ f_2^F \end{pmatrix} \quad \text{and} \quad h_2 = \begin{pmatrix} h_2^E \\ h_2^F \end{pmatrix}.$$

The first homological equation in (8.34) takes the form

$$\begin{cases} L_E h_2^E = f_2^E, \\ L_F h_2^F = f_2^F \end{cases} \quad \Leftrightarrow \quad \begin{cases} h_2^E = L_E^{-1} f_2^E, \\ L_F h_2^F = f_2^F. \end{cases}$$

Substituting $h_2^E$ in (8.31), we finally obtain

$$y' = Ay + f_2(y) - (L_A^{m,n}h_2)(y) + \cdots$$
$$= Ay + f_2^E(y) - (L_E h_2^E)(y) + f_2^F(y) - (L_F h_2^F)(y) \cdots$$
$$= Ay + f_2^F(y) - (L_F h_2^F)(y) + \cdots .$$

The terms of degree 2 with values in $F$ cannot be eliminated, unless we have already $f_2^F = 0$ from the beginning. Continuing this procedure for the remaining homological equations in (8.34), in general only the terms of the form (8.36) cannot be eliminated.

**Example 8.23.** Consider the equation

$$\begin{pmatrix} x \\ y \end{pmatrix}' = \begin{pmatrix} 0 & 2 \\ -2 & 0 \end{pmatrix} \begin{pmatrix} x \\ y \end{pmatrix} + \cdots . \tag{8.37}$$

The eigenvalues of the $2 \times 2$ matrix in (8.37) are $\pm 2i$, and thus, in order to determine whether there are resonant vectors we have to solve the equation

$$m_1 2i + m_2(-2i) = \pm 2i,$$

that is,

$$m_1 - m_2 = \pm 1,$$

with $m_1, m_2 \in \mathbb{N} \cup \{0\}$ and $m_1 + m_2 \geq 2$. The solutions are

$$(m_1, m_2) = (1, 2), (2, 1), (2, 3), (3, 2), \ldots .$$

By the former discussion, this implies that by a change of variables only the terms of the form

$$xy^2, \ x^2 y, \ x^2 y^3, \ x^3 y^2, \ \ldots$$

cannot be eliminated. In other words, denoting by $u$ and $v$ the components in the new coordinates, equation (8.37) takes the form

$$\begin{cases} u' = 2v + a_1 uv^2 + a_2 u^2 v + a_3 u^2 v^3 + a_4 u^3 v^2 + \cdots, \\ v' = -2u + b_1 uv^2 + b_2 u^2 v + b_3 u^2 v^3 + b_4 u^3 v^2 + \cdots \end{cases}$$

for some constants $a_i, b_i \in \mathbb{R}$ with $i \in \mathbb{N}$.

**Example 8.24.** Consider the equation

$$\begin{pmatrix} x \\ y \end{pmatrix}' = \begin{pmatrix} 2 & 0 \\ 0 & 1 \end{pmatrix} \begin{pmatrix} x \\ y \end{pmatrix} + \cdots . \tag{8.38}$$

The eigenvalues of the $2 \times 2$ matrix in (8.38) are 2 and 1, and thus, in order to determine whether there are resonant vectors we have to solve the equations

$$2m_1 + m_2 = 2 \quad \text{and} \quad 2m_1 + m_2 = 1,$$

with $m_1, m_2 \in \mathbb{N} \cup \{0\}$ and $m_1 + m_2 \geq 2$. One can easily verify that $(m_1, m_2) = (0, 2)$ is the only solution. This implies that denoting by $u$ and $v$

the components after an appropriate change of variables, equation (8.38) takes the form

$$\begin{cases} u' = 2u + \alpha v^2, \\ v' = v + \beta v^2 \end{cases}$$

for some constants $\alpha, \beta \in \mathbb{R}$. Moreover, one can take $\beta = 0$, since by Proposition 8.19 the solution $(m_1, m_2) = (0, 2)$ corresponds to the eigenvector $(y^2, 0)$ (see (8.36)), which has the second component equal to zero.

We conclude this section with a general result concerning the perturbations of linear equations whose matrices have only eigenvalues with positive real part.

**Theorem 8.25.** *Let $A$ be an $n \times n$ matrix having only eigenvalues with positive real part and let $f \colon \mathbb{R}^n \to \mathbb{R}^n$ be a function of class $C^\infty$ with $f(0) = 0$ and $d_0 f = 0$. Then for each $k \in \mathbb{N}$ the equation $x' = Ax + f(x)$ can be transformed by a change of variables $x = y + h(y)$ into*

$$y' = Ay + p(y) + o(\|y\|^k),$$

*where $p$ is a polynomial.*

**Proof.** In order to verify whether there are resonant vectors, consider the equation

$$\lambda_j = \sum_{i=1}^{n} m_i \lambda_i. \tag{8.39}$$

We note that

$$\operatorname{Re} \lambda_j = \sum_{i=1}^{n} m_i \operatorname{Re} \lambda_i. \tag{8.40}$$

Taking a vector $(m_1, \ldots, m_n)$ such that

$$\sum_{i=1}^{n} m_i > \frac{\max_i \operatorname{Re} \lambda_i}{\min_i \operatorname{Re} \lambda_i}$$

(we recall that by hypothesis the eigenvalues have positive real part), we obtain

$$\sum_{i=1}^{n} m_i \operatorname{Re} \lambda_i \geq \sum_{i=1}^{n} m_i \min_i \operatorname{Re} \lambda_i$$
$$> \max_i \operatorname{Re} \lambda_i \geq \operatorname{Re} \lambda_j.$$

This shows that identity (8.40), and thus also identity (8.39), do not hold for the vector $(m_1, \ldots, m_n)$. Therefore, there exist at most finitely many resonant vectors $(m_1, \ldots, m_n)$, taken among those such that

$$\sum_{i=1}^{n} m_i \le \frac{\max_i \operatorname{Re} \lambda_i}{\min_i \operatorname{Re} \lambda_i}.$$

This yields the desired result.                                                                $\square$

## 8.4. Exercises

**Exercise 8.1.** Consider the equation

$$\begin{cases} x' = xy + ax^3 + bxy^2, \\ y' = -y + cx^2 + dx^2y. \end{cases}$$

   a) Find a center manifold of the origin up to third order.

   b) Determine the stability of the origin when $a + c < 0$.

**Exercise 8.2.** Consider the equation

$$\begin{cases} x' = 2x + p(x, y), \\ y' = y + q(x, y), \end{cases}$$

where $p$ and $q$ are polynomials without terms of degrees 0 or 1.

   a) Find the stable, unstable and center manifolds of the origin.

   b) Show that by an appropriate change of variables the equation can be transformed into $z' = 2z + cw^2$, $w' = w$.

**Exercise 8.3.** For each $\varepsilon \in \mathbb{R}$, consider the equation

$$\begin{cases} x' = y - x^3, \\ y' = \varepsilon y - x^3. \end{cases}$$

   a) Use a Lyapunov function to show that the origin is stable when $\varepsilon = 0$.

   b) Find a center manifold of the origin up to fifth order.

   c) Determine the stability of the origin for each $\varepsilon \ne 0$.

   d) Find a center manifold of the origin when $\varepsilon = 1$.

   e) Find whether there exist heteroclinic orbits when $\varepsilon = 1$.

**Exercise 8.4.** Consider the equation

$$\begin{cases} x' = y + x^3, \\ y' = \varepsilon x - y^2. \end{cases}$$

   a) Find whether there exist periodic orbits contained in the first quadrant when $\varepsilon = 1$.

   b) Determine the stability of all critical points when $\varepsilon = 1$.

c) Determine the stability of the origin when $\varepsilon = 0$.

d) Find whether any bifurcation occurs for some $\varepsilon > 0$.

**Exercise 8.5.** For each $(\varepsilon, \delta) \in \mathbb{R}^2$, consider the equation

$$\begin{cases} x' = \varepsilon x - y + \delta x(x^2 + y^2), \\ y' = x - \varepsilon y + \delta y(x^2 + y^2). \end{cases}$$

a) Determine the stability of the origin for each pair $(\varepsilon, \delta)$ with $|\varepsilon| < 1$. Hint: For the function

$$V(x, y) = x^2 + y^2 - 2\varepsilon xy$$

we have

$$\dot{V}(x, y) = 2\delta(x^2 + y^2)V(x, y).$$

b) For each pair $(\varepsilon, \delta)$ with $|\varepsilon| > 1$, show that no bifurcations occur in a neighborhood of the origin.

c) For each pair $(\varepsilon, \delta)$ with $\delta < 0$, show that each positive semiorbit is bounded.

**Solutions.**

**8.1** a) $W^c = \{(x, cx^2 + o(x^3)) : x \in (-\delta, \delta)\}$.

b) The origin is unstable.

**8.2** a) $W^s = \{(0,0)\}$, $W^u = (-\delta, \delta) \times (-\delta, \delta)$ and $W^c = \{(0,0)\}$.

**8.3** a) $(x^4 + 2y^2)' = 4(x^3 x' + yy') = -4x^6 \le 0$.

b) $W^c = \begin{cases} (-\delta, \delta) \times (-\delta, \delta) & \text{if } \varepsilon = 0, \\ \{(x, x^3/\varepsilon + 3x^5/\varepsilon^3 + \cdots) : x \in (-\delta, \delta)\} & \text{if } \varepsilon \ne 0. \end{cases}$

c) Unstable.

d) $W^c = \{(x, x^3) : x \in (-\delta, \delta)\}$.

e) There exist.

**8.4** a) There are none.

b) $(0,0)$ are $(1, -1)$ are unstable.

c) Unstable.

d) There are no bifurcations.

**8.5** a) Unstable for $\delta > 0$, stable but not asymptotically stable for $\delta = 0$, and asymptotically stable for $\delta < 0$.

# Chapter 9

# Hamiltonian Systems

In this chapter we give a brief introduction to the theory of Hamiltonian systems. These are particularly important in view of their ubiquity in physical systems. After introducing the basic notions of the theory, we establish some results concerning the stability of linear and nonlinear Hamiltonian systems. We also consider the notion of integrability and, in particular, the Liouville–Arnold theorem on the structure of the level sets of independent integrals in involution. In addition, we briefly describe the basic ideas of the Kolmogorov–Arnold–Moser theory. For additional topics we refer the reader to [1, 3, 22, 25].

## 9.1. Basic notions

Let $H: \mathbb{R}^{2n} \to \mathbb{R}$ be a function of class $C^2$, called a *Hamiltonian*. More generally, one can consider Hamiltonians in an open subset of $\mathbb{R}^{2n}$. We write the coordinates of $\mathbb{R}^{2n}$ in the form $(q, p)$, where $q = (q_1, \ldots, q_n)$ and $p = (p_1, \ldots, p_n)$.

**Definition 9.1.** The system of equations

$$q_j' = \frac{\partial H}{\partial p_j}(q, p), \quad p_j' = -\frac{\partial H}{\partial q_j}(q, p), \quad j = 1, \ldots, n \tag{9.1}$$

is called a *Hamiltonian system* with *n degrees of freedom*.

The following are examples of Hamiltonian systems.

**Example 9.2.** Let $U: \mathbb{R} \to \mathbb{R}$ be a function of class $C^2$. Letting $q' = p$, the equation $q'' = -U'(q)$ is equivalent to the Hamiltonian system with one degree of freedom determined by the Hamiltonian

$$H(q,p) = \frac{p^2}{2} + U(q). \tag{9.2}$$

It is easy to verify that $H$ is constant along the solutions of the equation

$$\begin{cases} q' = p, \\ p' = -U'(q). \end{cases} \tag{9.3}$$

Incidentally, the terms $p^2/2$ and $U(q)$ in (9.2) correspond respectively to the kinetic and potential energies.

**Example 9.3.** The Kepler problem is a special case of the two-body problem and can be described by a Hamiltonian system with two degrees of freedom; namely, consider the Hamiltonian

$$H(q,p) = \frac{\|p\|^2}{2} - U(q) \tag{9.4}$$

in $\mathbb{R}^4$, for the potential $U(q) = -\mu/\|q\|$, with $\mu \in \mathbb{R}$. The corresponding Hamiltonian system is

$$q' = p, \quad p' = -\frac{\mu q}{\|q\|^3}. \tag{9.5}$$

Now we write system (9.1) in the form

$$x' = X_H(x) = J\nabla H(x), \tag{9.6}$$

where $x = (q,p) \in \mathbb{R}^{2n}$,

$$\nabla H = \left( \frac{\partial H}{\partial q_1}, \dots, \frac{\partial H}{\partial q_n}, \frac{\partial H}{\partial p_1}, \dots, \frac{\partial H}{\partial p_n} \right)$$

and

$$J = \begin{pmatrix} 0 & \mathrm{Id} \\ -\mathrm{Id} & 0 \end{pmatrix}.$$

Here $\mathrm{Id} \in M_n$ is the identity matrix. In particular, we would like to know which transformations preserve the Hamiltonian structure, in the following sense. Making the change of variables $x = \varphi(y)$, the Hamiltonian system (9.6) is transformed into

$$y' = (\varphi^* X_H)(y),$$

where

$$(\varphi^* X_H)(y) = (d_y \varphi)^{-1} X_H(\varphi(y)).$$

We would have $\varphi^* X_H = X_{H \circ \varphi}$, and hence also

$$y' = X_{H \circ \varphi}(y),$$

which would correspond to the preservation of the Hamiltonian structure, if and only if

$$(d_y \varphi)^{-1} J \nabla H(\varphi(y)) = J \nabla (H \circ \varphi)(y)$$

for every $y \in \mathbb{R}^{2n}$. Since

$$\nabla(H \circ \varphi)(y) = (d_y\varphi)^*\nabla H(\varphi(y)),$$

the identity $\varphi^*X_H = X_{H\circ\varphi}$ holds whenever

$$(d_y\varphi)^{-1}J = J(d_y\varphi)^*. \tag{9.7}$$

Since $J^{-1} = -J$, identity (9.7) can be written in the form

$$(d_y\varphi)^*Jd_y\varphi = J. \tag{9.8}$$

**Definition 9.4.** A differentiable transformation $\varphi\colon V \to \mathbb{R}^{2n}$ in an open set $V \subset \mathbb{R}^{2n}$ is said to be *canonical* if identity (9.8) holds for every $y \in V$.

**Example 9.5.** We show that for $n = 1$ a differentiable transformation $\varphi\colon V \to \mathbb{R}^2$ is canonical if and only if $\det d_y\varphi = 1$ for every $y \in V$. Writing

$$d_y\varphi = \begin{pmatrix} a & b \\ c & d \end{pmatrix},$$

identity (9.8) is equivalent to

$$\begin{pmatrix} a & b \\ c & d \end{pmatrix}^* \begin{pmatrix} 0 & 1 \\ -1 & 0 \end{pmatrix} \begin{pmatrix} a & b \\ c & d \end{pmatrix} = \begin{pmatrix} 0 & ad-bc \\ -ad+bc & 0 \end{pmatrix} = \begin{pmatrix} 0 & 1 \\ -1 & 0 \end{pmatrix}.$$

The last equality holds if and only if

$$\det d_y\varphi = ad - bc = 1.$$

This shows that for $n = 1$ the canonical transformations are the differentiable transformations preserving area and orientation.

Now we show that the solutions of a Hamiltonian system induce canonical transformations preserving volume and orientation. Let $H\colon \mathbb{R}^{2n} \to \mathbb{R}$ be a Hamiltonian of class $C^2$. Given $x \in \mathbb{R}^{2n}$, we denote by $\varphi_t(x)$ the solution of the Hamiltonian system (9.6) with initial condition $\varphi_0(x) = x$.

**Proposition 9.6.** *Given a Hamiltonian $H$ of class $C^2$, for each sufficiently small $t$ the map $\varphi_t$ is a canonical transformation preserving volume and orientation.*

**Proof.** We have

$$\frac{\partial}{\partial t}\varphi_t(x) = X_H(\varphi_t(x)). \tag{9.9}$$

Since $H$ is of class $C^2$, it follows from (9.9) and an appropriate generalization of Theorem 1.42 for $C^2$ vector fields that

$$\frac{\partial}{\partial t}d_x\varphi_t = d_{\varphi_t(x)}X_H d_x\varphi_t.$$

This shows that $d_x\varphi_t$ is a solution of the linear variational equation of the Hamiltonian system along the solution $\varphi_t(x)$. Moreover, since $\varphi_0(x) = x$, we have $d_x\varphi_0 = \text{Id}$. Now let

$$\alpha(t) = (d_x\varphi_t)^* J d_x\varphi_t.$$

We want to show that $\alpha(t) = J$. Taking derivatives with respect to $t$, we obtain

$$\alpha'(t) = (d_{\varphi_t(x)}X_H d_x\varphi_t)^* J d_x\varphi_t + (d_x\varphi_t)^* J d_{\varphi_t(x)}X_H d_x\varphi_t$$
$$= (d_x\varphi_t)^* \big[(d_{\varphi_t(x)}X_H)^* J + J d_{\varphi_t(x)}X_H\big] d_x\varphi_t,$$

and it follows from the identity $d_y X_H = J d_y^2 H$ that

$$\alpha'(t) = (d_x\varphi_t)^* \big(d_{\varphi_t(x)}^2 H J^* J + J^2 d_{\varphi_t(x)}^2 H\big) d_x\varphi_t.$$

Since $J^* J = \text{Id}$ and $J^2 = -\text{Id}$, we obtain

$$\alpha'(t) = (d_x\varphi_t)^* \big(d_{\varphi_t(x)}^2 H - d_{\varphi_t(x)}^2 H\big) d_x\varphi_t = 0.$$

On the other hand, since $d_x\varphi_0 = \text{Id}$, we have $\alpha(0) = J$ and thus $\alpha(t) = \alpha(0) = J$. This shows that $\varphi_t$ is a canonical transformation. Since

$$\det(d_x\varphi_t)^* = \det d_x\varphi_t,$$

it follows from identity (9.8) that $|\det d_x\varphi_t|^2 = 1$. But since $\det d_x\varphi_0 = \det \text{Id} = 1$, it follows from the continuity of $t \mapsto \varphi_t(x)$ that $\det d_x\varphi_t = 1$ for any sufficiently small $t$. This establishes the desired result. $\square$

**Example 9.7.** Consider the Hamiltonian system with one degree of freedom

$$\begin{cases} q' = p, \\ p' = -q, \end{cases}$$

determined by the Hamiltonian

$$H(q, p) = \frac{p^2}{2} + \frac{q^2}{2}.$$

By Example 1.8, we have

$$\varphi_t(q_0, p_0) = (p_0 \sin t + q_0 \cos t, p_0 \cos t - q_0 \sin t)$$
$$= \begin{pmatrix} \cos t & \sin t \\ -\sin t & \cos t \end{pmatrix} \begin{pmatrix} q_0 \\ p_0 \end{pmatrix},$$

and clearly,

$$\det d_{(q_0, p_0)}\varphi_t = \det \begin{pmatrix} \cos t & \sin t \\ -\sin t & \cos t \end{pmatrix} = 1.$$

We also note that for any differentiable function $F \colon \mathbb{R}^{2n} \to \mathbb{R}$, given a solution $x(t)$ of equation (9.6) we have

$$\frac{d}{dt}F(x(t)) = \nabla F(x(t)) \cdot X_H(x(t)) = (\nabla F(x(t)))^* J \nabla H(x(t)).$$

**Definition 9.8.** The *Poisson bracket* $\{F, G\}$ of two differentiable functions $F, G \colon \mathbb{R}^{2n} \to \mathbb{R}$ is defined by

$$\{F, G\}(x) = (\nabla F(x))^* J \nabla G(x) = \sum_{j=1}^{n} \left( \frac{\partial F}{\partial q_j}(x) \frac{\partial G}{\partial p_j}(x) - \frac{\partial F}{\partial p_j}(x) \frac{\partial G}{\partial q_j}(x) \right).$$

If $x(t)$ is a solution of the Hamiltonian system (9.6) and $F$ is a differentiable function, then

$$\frac{d}{dt} F(x(t)) = \{F, H\}(x(t)).$$

Therefore, in order that $F$ is an integral of equation (9.6) (see Definition 1.67) it is necessary that $\{F, H\} = 0$. Incidentally, we have $\{H, H\} = 0$, and thus any Hamiltonian is constant along the solutions of its Hamiltonian system.

## 9.2. Linear Hamiltonian systems

In this section we consider a particular class of Hamiltonians. Namely, given a $2n \times 2n$ symmetric matrix $S$, consider the Hamiltonian

$$H(x) = \frac{1}{2} x^* S x.$$

Clearly, $H$ is a polynomial of degree 2 without terms of degree 0 or 1. The corresponding Hamiltonian system, called a *linear Hamiltonian system*, is

$$x' = X_H(x) = JSx.$$

**Definition 9.9.** A matrix $B = JS$ with $S$ symmetric is said to be *Hamiltonian*. A matrix $A$ is said to be *canonical* if

$$A^* J A = J. \tag{9.10}$$

We give several examples.

**Example 9.10.** It follows from (9.8) that if $\varphi \colon V \to \mathbb{R}^{2n}$ is a canonical transformation, then $d_y \varphi$ is a canonical matrix for each $y \in V$.

**Example 9.11.** We show that a matrix $B$ is Hamiltonian if and only if

$$B^* J + J B = 0. \tag{9.11}$$

Indeed, if $B$ is Hamiltonian, then

$$B^* J + J B = S^* J^* J + J^2 S = S^* - S = 0.$$

On the other hand, if identity (9.11) holds, then

$$(-JB)^* = B^*(-J)^* = B^* J = -JB,$$

and the matrix $S = -JB$ is symmetric. Moreover, $JS = -J^2 B = B$ and the matrix $B$ is Hamiltonian.

**Example 9.12.** We show that any exponential $e^{Bt}$ of a Hamiltonian matrix $B$ is a canonical matrix. Let $B = JS$, with $S$ symmetric, be a Hamiltonian matrix, and consider the function

$$\beta(t) = (e^{Bt})^* J e^{Bt}.$$

We have $\beta(0) = J$ and

$$\begin{aligned}
\beta'(t) &= (JSe^{JSt})^* J e^{JSt} + (e^{JSt})^* J J S e^{JSt} \\
&= (e^{Bt})^* S^* J^* J e^{Bt} + (e^{Bt})^* J^2 S e^{Bt} \\
&= (e^{Bt})^* S e^{Bt} - (e^{Bt})^* S e^{Bt} = 0,
\end{aligned}$$

because the matrix $S$ is symmetric. This shows that $\beta(t) = J$ for $t \in \mathbb{R}$, and each matrix $e^{Bt}$ is canonical.

Now we describe some properties of the eigenvalues of Hamiltonian matrices and canonical matrices.

**Proposition 9.13.** *Let $B$ be a Hamiltonian matrix. If $\lambda$ is an eigenvalue of $B$, then $-\lambda$, $\bar{\lambda}$ and $-\bar{\lambda}$ are also eigenvalues of $B$, with the same multiplicity as $\lambda$. Moreover, if $0$ is an eigenvalue of $B$, then it has even multiplicity.*

**Proof.** It follows from (9.11) that $B = J^{-1}(-B^*)J$. Hence, the matrices $B$ and $-B^*$ are similar and they have the same eigenvalues (with the same multiplicity). Thus, if $\lambda$ is an eigenvalue of $B$, then $-\bar{\lambda}$ is an eigenvalue of $-B^*$ and hence also of $B$ (with the same multiplicity). Since $B$ is real, we conclude that $\bar{\lambda}$ and $-\lambda$ are also eigenvalues (with the same multiplicity). In particular, the nonzero eigenvalues of $B$ form groups of two or four, and thus if $0$ is an eigenvalue, then it has even multiplicity. $\square$

**Proposition 9.14.** *Let $A$ be a canonical matrix. If $\lambda$ is an eigenvalue of $A$, then $1/\lambda$, $\bar{\lambda}$ and $-1/\bar{\lambda}$ are also eigenvalues of $A$, with the same multiplicity as $\lambda$. Moreover, if $1$ and $-1$ are eigenvalues of $A$, then each of them has even multiplicity.*

**Proof.** It follows from (9.10) that $A$ is invertible and $A^* = JA^{-1}J^{-1}$. In particular, the matrices $A^{-1}$ and $A^*$ are similar and they have the same eigenvalues (with the same multiplicity). Thus, if $\lambda$ is an eigenvalue of $A$, then $1/\lambda$ is an eigenvalue of $A^{-1}$ and hence also of $A^*$. This shows that $1/\bar{\lambda}$ is an eigenvalue of $A$ (with the same multiplicity as $\lambda$). Since $A$ is real, we conclude that $\bar{\lambda}$ and $1/\lambda$ are also eigenvalues (with the same multiplicity). The last property follows immediately from the previous ones. $\square$

By Theorem 3.10, a linear equation $x' = Bx$ is asymptotically stable if and only if $B$ has only eigenvalues with negative real part, and is stable if and only if $B$ has no eigenvalues with positive real part and each eigenvalue with zero real part has a diagonal Jordan block. It follows easily from Proposition 9.13 that for Hamiltonian matrix $B$ the linear equation $x' = Bx$ is not asymptotically stable. Moreover, if the equation is stable, then all eigenvalues of $B$ are purely imaginary and have a diagonal Jordan block.

## 9.3. Stability of equilibria

In this section we study the stability of the critical points of a nonlinear Hamiltonian system. Let $H \colon \mathbb{R}^{2n} \to \mathbb{R}$ be a Hamiltonian of class $C^2$ and consider the corresponding Hamiltonian system $x' = X_H(x)$.

**Definition 9.15.** A point $x_0 \in \mathbb{R}^{2n}$ is said to be an *equilibrium* of the Hamiltonian $H$ if $\nabla H(x_0) = 0$. An equilibrium $x_0$ of $H$ is said to be *nondegenerate* if the Hessian matrix $d_{x_0}^2 H$ is nonsingular.

We note that the equilibria of $H$ are critical points of $x' = X_H(x)$.

We recall that the *index* of a bilinear transformation $F \colon \mathbb{R}^{2n} \times \mathbb{R}^{2n} \to \mathbb{R}$, such as, for example, $(u, v) \mapsto u^* d_{x_0}^2 H v$, is the maximal dimension of the subspaces $E \subset \mathbb{R}^{2n}$ such that $F$ is negative definite in $E \times E$. The following statement shows that the form of a Hamiltonian $H$ in a neighborhood of a nondegenerate equilibrium $x_0$ is completely determined by the index of the Hessian matrix $d_{x_0}^2 H$ (for a proof see, for example, [**21**]).

**Proposition 9.16** (Morse's lemma). *If $H \colon \mathbb{R}^{2n} \to \mathbb{R}$ is a function of class $C^2$ and $x_0 \in \mathbb{R}^{2n}$ is a nondegenerate equilibrium of $H$, then there exists a change of variables $g \colon B(0, r) \to V$, where $V$ is a neighborhood of $x_0$, such that $g(0) = x_0$ and*

$$
(H \circ g)(y_1, \dots, y_{2n}) = H(x_0) - \sum_{i=1}^{\lambda} y_i^2 + \sum_{i=\lambda+1}^{2n} y_i^2
$$

*for every $(y_1, \dots, y_{2n}) \in B(0, r)$, where $\lambda$ is the index of $d_{x_0}^2 H$.*

Now we present a first result on the stability of the equilibria of a nonlinear Hamiltonian system. The proof uses Morse's lemma.

**Theorem 9.17.** *Let $H$ be a Hamiltonian of class $C^2$. If $x_0$ is an equilibrium of $H$ with (positive or negative) definite Hessian matrix $d_{x_0}^2 H$, then $x_0$ is a stable but not asymptotically stable critical point of the equation $x' = X_H(x)$.*

**Proof.** By hypothesis, the index of the Hessian matrix $d_{x_0}^2 H$ is 0 or $2n$. Hence, it follows from Morse's lemma that there exists a change of variables $g \colon B(0, r) \to V$, where $V$ is a neighborhood of $x_0$, such that $g(0) = x_0$ and

$$
(H \circ g)(y_1, \dots, y_{2n}) = H(x_0) \pm \sum_{i=1}^{2n} y_i^2
$$

for every $(y_1, \ldots, y_{2n}) \in B(0, r)$. This shows that for $c$ sufficiently close to $H(x_0)$, with $c > H(x_0)$ for $d_{x_0}^2 H$ positive definite, and $c < H(x_0)$ for $d_{x_0}^2 H$ negative definite, the level sets

$$V_c = \{x \in V : H(x) = c\}$$

are diffeomorphic to $2n$-dimensional spheres that approach the point $x_0$ when $c \to H(x_0)$. On the other hand, since $H$ is constant along the solutions of its Hamiltonian system, any solution $\varphi_t(x)$ with initial condition $x \in V_c$ remains in $V_c$ for every $t \in \mathbb{R}$ (we note that since $V_c$ is compact, the maximal interval of any of these solutions is $\mathbb{R}$). This yields the desired result.     $\square$

**Example 9.18** (continuation of Example 9.2). Consider the Hamiltonian $H$ in (9.2). It follows from Theorem 9.17 that if $q_0$ is a strict local minimum of $U$ with $U''(q_0) > 0$, then $(q_0, 0)$ is a stable but not asymptotically stable critical point of equation (9.3). Indeed, since $q_0$ is a local minimum of $U$, we have $U'(q_0) = 0$ and hence $(q_0, 0)$ is a critical point of equation (9.3). Moreover, since $U''(q_0) > 0$, the Hessian matrix

$$d_{(q_0, 0)}^2 H = d_{(q_0, 0)}(U', p) = \begin{pmatrix} U''(q_0) & 0 \\ 0 & 1 \end{pmatrix}$$

is positive definite.

Now we consider the case when $d_{x_0}^2 H$ is not definite. In order to study the stability of the equilibrium $x_0$ it is convenient to consider the linear variational equation

$$y' = d_{x_0} X_H y.$$

**Definition 9.19.** The eigenvalues of the Hamiltonian matrix $d_{x_0} X_H = J d_{x_0}^2 H$ are called the *characteristic exponents* of $X_H$ at $x_0$.

It follows from Theorem 4.9 that if some characteristic exponent of $X_H$ at $x_0$ has positive real part, then $x_0$ is an unstable critical point of the equation $x' = X_H(x)$. Since the matrix $d_{x_0} X_H$ is Hamiltonian, it follows from Proposition 9.13 that a necessary condition for the stability of an equilibrium $x_0$ is that all characteristic exponents of $X_H$ at $x_0$ are purely imaginary, say equal to $\pm i\lambda_1, \ldots, \pm i\lambda_n$. Now we give a sufficient condition for stability.

**Proposition 9.20.** *Let $x_0$ be an equilibrium of $H$ with characteristic exponents*

$$\pm i\lambda_1, \ldots, \pm i\lambda_n.$$

*If $|\lambda_1|, \ldots, |\lambda_n|$ are nonzero and pairwise distinct, then $x_0$ is a stable but not asymptotically stable critical point of the equation $x' = X_H(x)$.*

**Proof.** Without loss of generality, we assume that

$$\lambda_1 > \cdots > \lambda_n > 0. \tag{9.12}$$

Now let $w_1, \ldots, w_n, \overline{w_1}, \ldots, \overline{w_n} \in \mathbb{C}^{2n} \setminus \{0\}$ be the eigenvectors associated respectively to the eigenvalues $i\lambda_1, \ldots, i\lambda_n, -i\lambda_1, \ldots, -i\lambda_n$ of the matrix $B = Jd_{x_0}^2 H$. We write

$$u_j = \operatorname{Re} w_j \quad \text{and} \quad v_j = \operatorname{Im} w_j.$$

One can easily verify that $u_1, \ldots, u_n, v_1, \ldots, v_n$ is a basis of $\mathbb{R}^{2n}$, and it follows from the identity $Bw_j = i\lambda_j w_j$ that

$$Bu_j = -\lambda_j v_j \quad \text{and} \quad Bv_j = \lambda_j u_j \tag{9.13}$$

for $j = 1, \ldots, n$ (because $B$ is real). We also assume, without loss of generality, that

$$u_j^* J v_j = 1 \quad \text{for} \quad j = 1, \ldots, n. \tag{9.14}$$

Now we consider the bilinear transformation $\langle u, v \rangle = u^* J v$ in $\mathbb{C}^{2n} \times \mathbb{C}^{2n}$. We have

$$\langle u, v \rangle = \langle u, v \rangle^* = v^* J^* u = -v^* J u = -\langle v, u \rangle \tag{9.15}$$

for $u, v \in \mathbb{R}^{2n}$. In particular, taking $u = v$, we obtain

$$\langle u_j, u_j \rangle = \langle v_j, v_j \rangle = 0 \quad \text{for} \quad j = 1, \ldots, n. \tag{9.16}$$

Moreover, it follows from (9.14) and (9.15) that

$$\langle u_j, v_j \rangle = -\langle v_j, u_j \rangle = 1 \quad \text{for} \quad j = 1, \ldots, n. \tag{9.17}$$

We also have

$$\langle Bw_j, w_k \rangle = i\lambda_j \langle w_j, w_k \rangle, \tag{9.18}$$

and using (9.11), we obtain

$$\langle Bw_j, w_k \rangle = w_j^* B^* J w_k = -w_j^* J B w_k = -i\lambda_k \langle w_j, w_k \rangle. \tag{9.19}$$

Thus,

$$i(\lambda_j + \lambda_k)\langle w_j, w_k \rangle = 0,$$

and since $\lambda_j + \lambda_k > 0$, we conclude that

$$\langle w_j, w_k \rangle = 0 \quad \text{for} \quad j, k = 1, \ldots, n.$$

Substituting $w_j = u_j + iv_j$ and $w_k = u_k + iv_k$ yields the identity

$$\langle u_j, u_k \rangle + i\langle u_j, v_k \rangle + i\langle v_j, u_k \rangle - \langle v_j, v_k \rangle = 0. \tag{9.20}$$

Proceeding as in (9.18) and (9.19) with $w_j$ replaced by $\overline{w_j}$, we obtain

$$i(-\lambda_j + \lambda_k)\langle \overline{w_j}, w_k \rangle = 0 \quad \text{for} \quad j, k = 1, \ldots, n.$$

Thus, it follows from (9.12) that $\langle \overline{w_j}, w_k \rangle = 0$ for $j \neq k$, and hence,

$$\langle u_j, u_k \rangle + i\langle u_j, v_k \rangle - i\langle v_j, u_k \rangle + \langle v_j, v_k \rangle = 0 \tag{9.21}$$

for $j \neq k$. Adding and subtracting (9.20) and (9.21) yields the identities

$$\langle u_j, u_k \rangle + i\langle u_j, v_k \rangle = 0$$

and

$$i\langle v_j, u_k\rangle - \langle v_j, v_k\rangle = 0$$

for $j \neq k$. Taking the real and imaginary parts, we finally obtain

$$\langle u_j, u_k\rangle = \langle u_j, v_k\rangle = \langle v_j, u_k\rangle = \langle v_j, v_k\rangle = 0 \qquad (9.22)$$

for $j \neq k$. It follows from identities (9.16), (9.17) and (9.22) that the matrix $C$ with columns $u_1, \ldots, u_n, v_1, \ldots, v_n$ satisfies

$$C^*JC = J.$$

In other words, $C$ is a canonical matrix. Moreover, by (9.13), we have

$$C^{-1}BC = \begin{pmatrix} & & & \lambda_1 & & \\ & & & & \ddots & \\ & & & & & \lambda_n \\ -\lambda_1 & & & & & \\ & \ddots & & & & \\ & & -\lambda_n & & & \end{pmatrix},$$

and hence,

$$C^*JBC = JC^{-1}BC = -\begin{pmatrix} \lambda_1 & & & & & \\ & \ddots & & & & \\ & & \lambda_n & & & \\ & & & \lambda_1 & & \\ & & & & \ddots & \\ & & & & & \lambda_n \end{pmatrix}.$$

Therefore, the change of variables $x - x_0 = Cy$ transforms the Hamiltonian $H$ into the function

$$\begin{aligned} y \mapsto H(x_0) &+ \frac{1}{2}(Cy)^* d^2_{x_0} H Cy + o(\|y\|^2) \\ &= H(x_0) - \frac{1}{2}y^* C^* J B C y + o(\|y\|^2) \\ &= H(x_0) - \frac{1}{2}y^* J C^{-1} B C y + o(\|y\|^2) \\ &= H(x_0) + \frac{1}{2}\sum_{j=1}^{n} \lambda_j(\bar{q}_j^2 + \bar{p}_j^2) + o(\|y\|^2), \end{aligned}$$

where $y = (\bar{q}_1, \ldots, \bar{q}_n, \bar{p}_1, \ldots, \bar{p}_n)$. The desired result can now be obtained as in the proof of Theorem 9.17.                              $\square$

## 9.4. Integrability and action-angle coordinates

The solutions of a Hamiltonian system rarely can be obtained explicitly. In this section we consider a class of Hamiltonians for which this is possible. Let $H\colon \mathbb{R}^{2n} \to \mathbb{R}$ be a Hamiltonian of class $C^2$.

**Definition 9.21.** If there exists a canonical transformation into variables $(\theta, I) \in \mathbb{T}^n \times \mathbb{R}^n$ such that the Hamiltonian $H$ can be written in the form $H(q,p) = h(I)$, then the variables $(\theta, I)$ are called *action-angle coordinates*.

In action-angle coordinates, the Hamiltonian system can be written in the form
$$\theta' = \omega(I), \quad I' = 0,$$
where $\omega(I) = \nabla h(I)$. Hence, the solutions are given explicitly by
$$\theta(t) = \theta_0 + t\omega(I_0), \quad I(t) = I_0,$$
where $\theta_0 = \theta(0)$ and $I_0 = I(0)$.

**Example 9.22.** Consider the Hamiltonian
$$H(q,p) = \frac{1}{2} \sum_{j=1}^{n} \lambda_j(q_j^2 + p_j^2), \quad \lambda_j \in \mathbb{R} \setminus \{0\}. \tag{9.23}$$

One can pass to action-angle coordinates by applying the transformation
$$q_j = \sqrt{2I_j}\cos\theta_j, \quad p_j = -\sqrt{2I_j}\sin\theta_j, \quad j = 1,\ldots,n.$$

By Example 9.5, this transformation is canonical, because the derivative of $(q_j, p_j) \mapsto (\theta_j, I_j)$ has determinant 1 for $j = 1, \ldots, n$. In the new coordinates $(\theta, I)$ the Hamiltonian $\tilde{H}(\theta, I) = H(q,p)$ is given by
$$\tilde{H}(\theta, I) = \sum_{j=1}^{n} \lambda_j I_j.$$

The corresponding Hamiltonian system is
$$I_j' = 0, \quad \theta_j' = \lambda_j, \quad j = 1, \ldots, n.$$

In general, a Hamiltonian cannot be written in action-angle coordinates. However, there is a class of Hamiltonians, called completely integrable (or integrable in the sense of Liouville), for which locally there exist action-angle coordinates. More precisely, for a Hamiltonian $H\colon \mathbb{R}^{2n} \to \mathbb{R}$ we describe how the existence of $n$ integrals $F_1, \ldots, F_n$ satisfying certain properties in a set $U \subset \mathbb{R}^{2n}$ has important consequences concerning the existence of local action-angle coordinates and the structure of the solutions in each level set
$$M_c = \big\{ x \in U : F_j(x) = c_j \text{ for } j = 1, \ldots, n \big\}, \tag{9.24}$$
where $c = (c_1, \ldots, c_n) \in \mathbb{R}^n$. We first introduce some notions.

**Definition 9.23.** Given $n$ differentiable functions $F_1, \ldots, F_n \colon \mathbb{R}^{2n} \to \mathbb{R}$, we say that they are:

    a) *independent* in $U \subset \mathbb{R}^{2n}$ if $\nabla F_1(x), \ldots, \nabla F_n(x)$ are linearly independent for each $x \in U$;

    b) in *involution* in $U \subset \mathbb{R}^{2n}$ if $\{F_i, F_j\} = 0$ in $U$ for $i, j = 1, \ldots, n$.

We also introduce the notion of complete integrability.

**Definition 9.24.** A Hamiltonian $H$ is said to be *completely integrable* (or *integrable in the sense of Liouville*) in a set $U \subset \mathbb{R}^{2n}$ if there exist $n$ integrals $F_1, \ldots, F_n$ of equation (9.6) that are independent and in involution in $U$.

**Example 9.25** (continuation of Example 9.22)**.** Consider the Hamiltonian $H$ in (9.23). One can easily verify that the functions

$$H_j(q, p) = \lambda_j(q_j^2 + p_j^2)/2,$$

for $j = 1, \ldots, n$, are integrals, and that they are independent and in involution in $\mathbb{R}^{2n}$. Hence, the Hamiltonian $H$ is completely integrable.

**Example 9.26** (continuation of Example 9.3)**.** Consider the Hamiltonian $H$ in (9.4) and the angular momentum

$$L(q, p) = q_1 p_2 - q_2 p_1. \tag{9.25}$$

One can easily verify that $L$ is an integral. Indeed, it follows from (9.5) that if $x(t) = (q(t), p(t))$ is a solution, then

$$\frac{d}{dt} L(x(t)) = q_1' p_2 + q_1 p_2' - q_2' p_1 - q_2 p_1'$$

$$= p_1 p_2 - q_1 \frac{\mu q_2}{\|q\|^3} - p_2 p_1 + q_2 \frac{\mu q_1}{\|q\|^3} = 0.$$

Moreover, the integrals $H$ and $L$ are independent outside the origin and $\{H, L\} = 0$. For the first property, we observe that

$$\nabla H = \left( p, \frac{\mu q}{\|q\|^3} \right) \quad \text{and} \quad \nabla L = (p_2, -p_1, -q_2, q_1).$$

Hence $\nabla H \cdot \nabla L = 0$, and $\nabla H$ and $\nabla L$ are linearly independent outside the origin. For the second property, we note that

$$\{H, L\} = \frac{\partial H}{\partial q_1} \cdot \frac{\partial L}{\partial p_1} + \frac{\partial H}{\partial q_2} \cdot \frac{\partial L}{\partial p_2} - \frac{\partial H}{\partial p_1} \cdot \frac{\partial L}{\partial q_1} - \frac{\partial H}{\partial p_2} \cdot \frac{\partial L}{\partial q_2}$$

$$= \frac{\mu q_1}{\|q\|^3}(-q_2) + \frac{\mu q_2}{\|q\|^3} q_1 - p_1 p_2 - p_2(-p_1) = 0.$$

Hence, the Hamiltonian in (9.4) is completely integrable outside the origin.

Now we formulate the following important result without proof (for details see, for example, [**19**]). We denote by $\mathbb{T}^n = \mathbb{R}^n/\mathbb{Z}^n$ the $n$-dimensional torus.

**Theorem 9.27** (Liouville–Arnold). *For a Hamiltonian $H$ that is completely integrable in an open set $U \subset \mathbb{R}^{2n}$, let $F_1, \ldots, F_n\colon U \to \mathbb{R}$ be integrals of equation (9.6) that are independent and in involution in $U$. Then:*

a) *each set $M_c$ in (9.24) is an $n$-dimensional manifold that is invariant under the solutions of the equations $x' = X_{F_i}(x)$ for $i = 1, \ldots, n$;*

b) *if all solutions of these equations in $M_c$ have maximal interval $\mathbb{R}$, then each connected component of $M_c$ is diffeomorphic to $\mathbb{T}^k \times \mathbb{R}^{n-k}$ for some $0 \le k \le n$, and there exist action-angle coordinates in a neighborhood of $M_c$; moreover, the solutions of the equation $x' = X_H(x)$ in $M_c$ induce the trajectories*

$$t \mapsto (\varphi_0 + t\omega \bmod 1, y_0 + t\nu)$$

*in $\mathbb{T}^k \times \mathbb{R}^{n-k}$, where $\omega = \omega(c) \in \mathbb{R}^k$ and $\nu = \nu(c) \in \mathbb{R}^{n-k}$.*

We first consider an example in $\mathbb{R}^2$.

**Example 9.28.** The Hamiltonian $H$ in (9.2) is completely integrable in a neighborhood of any point $(q_0, 0) \in \mathbb{R}^2$ where $q_0$ is a strict local minimum of $U$. Moreover, by Morse's lemma (Proposition 9.16), up to a change of variables in a neighborhood of $(q_0, 0)$ the level sets $M_c$ in (9.24) are circles. This corresponds to taking $n = k = 1$ in Theorem 9.27, in which case the sets $M_c$ are diffeomorphic to the torus $\mathbb{T}^1 = S^1$.

Now we present an example illustrating the complexity that can occur in the decomposition into level sets in Theorem 9.27. The details fall outside the scope of the book, and so we partially omit them.

**Example 9.29** (continuation of Example 9.26). Consider the Hamiltonian $H$ in (9.4) and the angular momentum $L$ in (9.25). Introducing the canonical transformation

$$r = \sqrt{q_1^2 + q_2^2}, \quad \theta = \arctan \frac{q_2}{q_1}, \quad p_r = \frac{q_1 p_1 + q_2 p_2}{\sqrt{q_1^2 + q_2^2}}, \quad p_\theta = q_1 p_2 - q_2 p_1,$$

one can write

$$H = \frac{1}{2}\left( p_r^2 + \frac{p_\theta^2}{r^2} \right) - \frac{\mu}{r} \quad \text{and} \quad L = p_\theta.$$

The Hamiltonian system can now be written in the form

$$r' = \frac{\partial H}{\partial p_r}, \quad \theta' = \frac{\partial H}{\partial p_\theta}, \quad p_r' = -\frac{\partial H}{\partial r}, \quad p_\theta' = -\frac{\partial H}{\partial \theta},$$

that is,

$$\begin{cases} r' = p_r, \\ \theta' = p_\theta/r^2, \\ p_r' = -p_\theta^2/r^3 + \mu/r^2, \\ p_\theta' = 0. \end{cases}$$

For the integrals $F_1 = H$ and $F_2 = L$, we have the level sets

$$M_c = \left\{ (r, \theta, p_r, p_\theta) \in \mathbb{R}^+ \times S^1 \times \mathbb{R}^2 : H(r, p_r, p_\theta) = a, p_\theta - b \right\}$$

$$= \left\{ (r, \theta, p_r, b) \in \mathbb{R}^+ \times S^1 \times \mathbb{R}^2 : \frac{1}{2}\left( p_r^2 + \frac{b^2}{r^2} \right) - \frac{\mu}{r} = a \right\},$$

where $c = (a, b)$. Since $p_r^2 + b^2/r^2 \geq 0$, the region in the variables $(r, \theta)$ where there exist solutions is given by

$$R_a = \left\{ (r, \theta) \in \mathbb{R}^+ \times S^1 : \frac{\mu}{r} \geq -a \right\}.$$

Using the symbol $A \approx B$ to indicate the existence of a diffeomorphism between $A$ and $B$, one can show that:

a) if $\mu < 0$, then $R_a = \varnothing$ for $a \leq 0$, and $R_a \approx (-\mu/a, +\infty) \times S^1$ for $a > 0$;

b) if $\mu = 0$, then $R_a = \varnothing$ for $a < 0$, and $R_a \approx \mathbb{R}^+ \times S^1$ for $a \geq 0$;

c) if $\mu > 0$, then $R_a \approx \mathbb{R}^+ \times S^1$ for $a \geq 0$, and $R_a \approx (0, -\mu/a) \times S^1$ for $a < 0$.

Now we define

$$S_a = \bigcup_{(r,\theta) \in R_a} I_a(r, \theta),$$

where

$$I_a(r, \theta) = \left\{ (r, \theta, p_r, p_\theta) \in \mathbb{R}^+ \times S^1 \times \mathbb{R}^2 : p_r^2 + p_\theta^2/r^2 = 2\left( \frac{\mu}{r} + a \right) \right\}.$$

The set $I_a(r, \theta)$ is an ellipse, a point or the empty set, respectively, if $(r, \theta)$ is in the interior, in the boundary or in the exterior of $R_a$. One can show that:

a) if $\mu > 0$, then $S_a \approx S^3 \setminus S^1$ for $a < 0$, and $S_a \approx S^3 \setminus (S^1 \cup S^1)$ for $a \geq 0$;

b) if $\mu = 0$, then $S_a = \varnothing$ for $a < 0$, and $S_a \approx S^3 \setminus (S^1 \cup S^1)$ for $a \geq 0$;

c) if $\mu < 0$, then $S_a = \varnothing$ for $a \leq 0$, and $S_a \approx S^3 \setminus S^1$ for $a > 0$.

Taking into account that

$$M_c = S_a \cap (\mathbb{R}^+ \times S^1 \times \mathbb{R} \times \{b\}),$$

one can then describe each level set up to a diffeomorphism.

An interesting particular case of the Liouville–Arnold theorem (Theorem 9.27) occurs when the level sets $M_c$ are compact. In this case each connected component of $M_c$ is diffeomorphic to the torus $\mathbb{T}^n$ (because $\mathbb{R}^m$ is not compact). Moreover, there is no need to assume as in Theorem 9.27 that the solutions of the equations $x' = X_{F_i}(x)$ have maximal interval $\mathbb{R}$, since then this property is automatically satisfied. In this context we introduce the following notion.

**Definition 9.30.** The trajectories in a torus $\mathbb{T}^n$ diffeomorphic to a level set $M_c$, given by

$$t \mapsto (\alpha_1 + t\omega_1, \ldots, \alpha_n + t\omega_n) \bmod 1, \tag{9.26}$$

are called *quasi-periodic trajectories* with *frequency vector* $\omega = (\omega_1, \ldots, \omega_n)$. The vector $\omega$ is said to be *nonresonant* if $\langle k, \omega \rangle \neq 0$ for every $k \in \mathbb{Z}^n \setminus \{0\}$.

One can easily obtain the following result.

**Proposition 9.31.** *Let $H$ be a completely integrable Hamiltonian. For each level set $M_c$ diffeomorphic to a torus $\mathbb{T}^n$, if $\omega$ is nonresonant, then each trajectory in (9.26) is dense in the torus.*

## 9.5. The KAM theorem

Although the completely integrable Hamiltonian systems are relatively rare, there are still many examples of Hamiltonian systems that are obtained from small perturbations of Hamiltonians of the form $h(I)$, using action-angle coordinates. A concrete example is given by the motion of the planets in the solar system, which is close to the problem of motion without taking into account the interactions between planets (but only between the Sun and each planet). This last system is completely integrable, since it consists of several two-body problems (see Examples 9.3 and 9.26).

In this context it is natural to introduce the following notion. We recall that for a function $g \colon \mathbb{R} \to \mathbb{R}$ we write $g(\varepsilon) = O(\varepsilon)$ if there exists a constant $C > 0$ such that $|g(\varepsilon)| \leq C|\varepsilon|$ for every $\varepsilon \in \mathbb{R}$.

**Definition 9.32.** A parameterized family of Hamiltonians $H_\varepsilon = H_\varepsilon(\theta, I)$, with $(\theta, I) \in \mathbb{T}^n \times G$ for some $G \subset \mathbb{R}^n$, is said to be *almost-integrable* if it is of the form

$$H_\varepsilon(\theta, I) = h(I) + f_\varepsilon(\theta, I), \quad \text{with} \quad f_\varepsilon(\theta, I) = O(\varepsilon), \tag{9.27}$$

for every $\varepsilon \in \mathbb{R}$ in some neighborhood of zero.

Let $H_\varepsilon$ be an almost-integrable family of Hamiltonians. The Hamiltonian system defined by $H_\varepsilon$ is

$$\theta' = \omega(I) + \frac{\partial f_\varepsilon}{\partial I}(\theta, I), \quad I' = -\frac{\partial f_\varepsilon}{\partial \theta}(\theta, I),$$

where $\omega(I) = \nabla h(I)$. For $\varepsilon = 0$ all $n$-dimensional tori $\mathbb{T}_I = \mathbb{T}^n \times \{I\}$ with $I \in G$ are invariant. Now we assume that

$$\det d_I^2 h \neq 0 \quad \text{for} \quad I \in G. \tag{9.28}$$

This condition guarantees that the transformation $I \mapsto \omega(I)$ is a local diffeomorphism, and hence all tori $\mathbb{T}_I$ can be (locally) parameterized by the components of the vector $\omega(I)$. We have the following fundamental result (for details see, for example, [22]).

**Theorem 9.33** (KAM theorem). *Let $H_\varepsilon$ be an analytic Hamiltonian in $\mathbb{T}^n \times G$ satisfying (9.27) and (9.28). Given $\tau > n-1$ and $\gamma > 0$, there exists $\varepsilon_0 = O(\gamma^2)$ such that if $|\varepsilon| \leq \varepsilon_0$, then for each torus $\mathbb{T}_I = \mathbb{T}^n \times \{I\}$ with*

$$|\langle k, \omega(I) \rangle| \geq \frac{\gamma}{\|k\|^\tau}, \quad k \in \mathbb{Z}^n \setminus \{0\}$$

*there exists an invariant torus $\mathbb{T}_I^\varepsilon$ of the Hamiltonian $H_\varepsilon$, with the same frequency vector $\omega(I)$, such that the maximal distance from $\mathbb{T}_I^\varepsilon$ to $\mathbb{T}_I$ is $O(\varepsilon/\gamma)$. Moreover, the volume of the complement in $\mathbb{T}^n \times G$ of the set covered by the tori $\mathbb{T}_I^\varepsilon$ is $O(\gamma)$.*

Theorem 9.33 says that under the nondegeneracy condition in (9.28), if $\varepsilon$ is sufficiently small, then many of the $n$-dimensional invariant tori are preserved in the perturbed system, up to a small deformation. When $n = 2$ one can deduce from Theorem 9.33 that the perturbed system has bounded trajectories, because the tori (which in this case are closed curves) separate the phase space (by Jordan's curve theorem). For $n \geq 3$, in general this property may not hold.

## 9.6. Exercises

**Exercise 9.1.** Show that a differentiable transformation $\varphi$ is canonical if and only if

$$\{F \circ \varphi, G \circ \varphi\} = \{F, G\} \circ \varphi$$

for any differentiable functions $F$ and $G$.

**Exercise 9.2.** For the flow $\varphi_t$ defined by a Hamiltonian system, use Liouville's formula (Theorem 2.10) to show that $\det d_x \varphi_t = 1$

**Exercise 9.3.** Show that:

   a) if a matrix $B$ is Hamiltonian, then $\operatorname{tr} B = 0$;

   b) if a matrix $A$ is canonical, then $\det A = 1$;

   c) for $n = 1$ a matrix $B$ is Hamiltonian if and only if $\operatorname{tr} B = 0$;

   d) for $n = 1$ a matrix $A$ is canonical if and only if $\det A = 1$.

**Exercise 9.4.** Verify that the equation $x'' + \sin x = 0$ is a Hamiltonian system with one degree of freedom.

**Exercise 9.5.** Let $H \colon \mathbb{R}^{2n} \to \mathbb{R}$ be a Hamiltonian of class $C^2$ such that the corresponding Hamiltonian system defines a flow $\varphi_t$ in $\mathbb{R}^{2n}$. Show that:

a) $H$ is constant along the solutions;

b) $\varphi_t$ preserves volume in $\mathbb{R}^{2n}$, that is, $\mu(\varphi_t(A)) = \mu(A)$ for every open set $A \subset \mathbb{R}^{2n}$ and $t \in \mathbb{R}$, where $\mu$ denotes the volume in $\mathbb{R}^{2n}$;

c) there exist neither asymptotically stable critical points nor asymptotically stable periodic solutions.

**Exercise 9.6.** For a function $f \colon \mathbb{R}^n \to \mathbb{R}^n$ of class $C^1$, assume that the equation $x' = f(x)$ defines a flow $\varphi_t$ preserving volume in $\mathbb{R}^n$ and that there exists a bounded set $A \subset \mathbb{R}^n$ such that $\varphi_t(A) = A$ for every $t \in \mathbb{R}$.

a) Given a nonempty open set $U \subset A$, show that not all sets $\varphi_1(U)$, $\varphi_2(U)$, $\varphi_3(U), \dots$ are pairwise disjoint.

b) Show that for each nonempty open set $U \subset A$ there exist $x \in U$ and $n \in \mathbb{N}$ such that $\varphi_{2n}(x) \in U$.

# Bibliography

[1] R. Abraham and J. Marsden, *Foundations of Mechanics*, Benjamin/Cummings, 1978.

[2] H. Amann, *Ordinary Differential Equations*, Walter de Gruyter, 1990.

[3] V. Arnold, *Mathematical Methods of Classical Mechanics*, Springer, 1989.

[4] V. Arnold, *Ordinary Differential Equations*, Springer, 1992.

[5] R. Bhatia, *Matrix Analysis*, Springer, 1997.

[6] J. Carr, *Applications of Centre Manifold Theory*, Springer, 1981.

[7] C. Chicone, *Ordinary Differential Equations with Applications*, Springer, 1999.

[8] S.-N. Chow and J. Hale, *Methods of Bifurcation Theory*, Springer, 1982.

[9] E. Coddington and N. Levinson, *Theory of Ordinary Differential Equations*, McGraw-Hill, 1955.

[10] T. Dieck, *Algebraic Topology*, European Mathematical Society, 2008.

[11] B. Dubrovin, A. Fomenko and S. Novikov, *Modern Geometry - Methods and Applications: The Geometry and Topology of Manifolds*, Springer, 1985.

[12] J. Guckenheimer and P. Holmes, *Nonlinear Oscillations, Dynamical Systems, and Bifurcations of Vector Fields*, Springer, 1983.

[13] J. Hale, *Ordinary Differential Equations*, Robert E. Krieger, 1980.

[14] J. Hale and H. Koçak, *Dynamics and Bifurcations*, Springer, 1991.

[15] P. Hartman, *Ordinary Differential Equations*, SIAM, 2002.

[16] N. Higham, *Functions of Matrices: Theory and Computation*, SIAM, 2008.

[17] M. Hirsch and S. Smale, *Differential Equations, Dynamical Systems, and Linear Algebra*, Academic Press, 1974.

[18] M. Irwin, *Smooth Dynamical Systems*, Academic Press, 1980.

[19] A. Katok and B. Hasselblatt, *Introduction to the Modern Theory of Dynamical Systems*, Cambridge University Press, 1995.

[20] J. La Salle and S. Lefschetz, *Stability by Liapunov's Direct Method, With Applications*, Academic Press, 1961.

[21] J. Milnor, *Morse Theory*, Princeton University Press, 1963.

[22] J. Moser, *Stable and Random Motions in Dynamical Systems*, Princeton University Press, 1973.

[23] J. Palis and W. de Melo, *Geometric Theory of Dynamical Systems*, Springer, 1982.

[24] D. Sánchez, *Ordinary Differential Equations and Stability Theory*, Dover, 1968.

[25] F. Verhulst, *Nonlinear Differential Equations and Dynamical Systems*, Springer, 1996.

# Index

# Selected Titles in This Series

# SELECTED TITLES IN THIS SERIES

For a complete list of titles in this series, visit the
AMS Bookstore at **www.ams.org/bookstore/**.